Infrastructural Ecologies

Infrastructural Ecologies

Alternative Development Models for Emerging Economies

Hillary Brown and Byron Stigge

The MIT Press
Cambridge, Massachusetts
London, England

This book was set in ITC Stone Serif Medium by Westchester Publishing Services.

Library of Congress Cataloging-in-Publication Data

Names: Brown, Hillary, author. | Stigge, Byron, author.
Title: Infrastructural ecologies : alternative development models for emerging economies / Hillary Brown and Byron Stigge.
Description: Cambridge, MA : MIT Press, 2017. | Includes bibliographical references and index.
Identifiers: LCCN 2016043789 | ISBN 9780262036337 (hardcover : alk. paper) | ISBN 9780262533867 (pbk. : alk. paper)
Subjects: LCSH: Infrastructure (Economics)—Developing countries. | Infrastructure (Economics)—Environmental aspects—Developing countries. | Economic development—Environmental aspects—Developing countries.
Classification: LCC HC59.72.C3 B76 2017 | DDC 338.9001—dc23
LC record available at https://lccn.loc.gov/2016043789

Contents

Acknowledgments   vii

1 Introduction: Closing the Infrastructure Gap   1

Learning from Caracol, Haiti   2

The Promise of Infrastructural Ecology   4

The Roots of the Infrastructure Gap   5

Infrastructural Ecology: Why and How   11

Industrial Symbiosis as a Model for Infrastructural Ecology: Two Examples   12

The Organization of This Book: The Five Objectives of Infrastructural Ecology   16

Imperatives for Infrastructural Ecologies   18

2 Solving for Pattern: From Interconnected to Symbiotic Systems   19

Preindustrial Ingenuity: Multifunctional River Crossings and Agro-Infrastructure   21

Simple Integration: Colocated Systems   25

Commensalist Associations   29

Reciprocity across Service Sectors   35

Integrating Multiple Systems: Toward a Circular Economy   40

Forward Thinking   46

3 The Soft Path: Aligning Water Infrastructure with Natural Systems   47

Multiple-Use Water Systems   48

Capture and Storage for Water Sufficiency   52

Green Infrastructure at Work in Emerging Economies   60

Water Reuse and Nutrient Recovery: Sustainable Imperatives for the Anthropocene   69

Heading Down the Soft Path   75

4   Post-Carbon Infrastructure: Power, Heat, and Transport   77

Emerging Economies and the Carbon Challenge   77

Alternative Power Production   80

Alternative Heat Production   92

Managing Waste for Energy   95

Decarbonizing Transportation   107

Low-Carbon Paths Forward   113

5   Climate-Adaptive Infrastructure: Responding to Changing Conditions   115

Coastal Protection and Adaptation: Hard and Soft Strategies   116

Inland Adaptations   128

Cross-Sector Solutions for Water Security   132

Looking Ahead: Climate and Infrastructural Ecologies   144

6   Infrastructural Coproduction: Inclusionary and Participatory Development   147

Decentralization and Community-Based Participation: Moving beyond Tokenism   151

Partnering for Service Provision   161

Entrepreneurship and Comprehensive Citizen Control   169

Stepping Up the Ladder   174

7   Implementing Infrastructural Ecologies: Improving the Odds   175

"How Are We Going to Pay for That?"   176

"Too Slow and Not Our Scope"   181

"That's Not How We Do It Here"   186

"Will the Next Administration Support This?"   189

Ways Forward   192

8   Putting the Five Objectives into Practice   193

Objective 1: Relational Solutions   193

Objective 2: Ecological Alignments   196

Objective 3: Low-Carbon Processes   199

Objective 4: Resilient Constructions   202

Objective 5: Codevelopment   205

Haiti Redux: A "Future-Proof" Vision?   207

Conclusion   211

Summary of Case Study Infrastructural Ecologies   213

Notes   235

Acronyms   281

Glossary   285

Recommended Readings   293

Index   297

Acknowledgments

The authors wish to express their gratitude to several colleagues and assistants who supported their work on this book. Special thanks go to Logman Arja, Benjamin Leer, Nehleen Chowdhury, Alexis Obernauer, April Schneider, Jorge Juan Cornet, and Rachel Galowich for their efforts, including case study research, graphic development, and other general assistance. The work has received support through the generosity of the Bernard and Anne Spitzer School of Architecture Fund.

The book was enhanced greatly by Sandra Chizinsky's keen editorial insights, organizing and clarifying the text to amalgamate two distinctive writing styles into a single coherent manuscript. Throughout the process, MIT Press's Environmental Studies acquisitions editor Beth Clevenger guided us patiently, rendering assistance with great care.

1

Introduction: Closing the Infrastructure Gap

The 145 nations that the United Nations classifies as "developing" represent more than six billion people, representing over 85 percent of the world's inhabitants but only 40 percent of global economic output.[1] Compared with so-called advanced economies, emerging economies have lower gross domestic product per capita, higher levels of unemployment and informal employment, and higher rates of poverty (both urban and rural). Most important for our purposes here, many emerging economies lack critical infrastructural services. According to 2011 projections, current infrastructure expenditures in less developed and least developed nations—which were just under US $1 trillion annually—must effectively double through 2020 in order to meet basic needs.[2] (Note: All dollar figures in this book are U.S. dollars unless otherwise stated.) Alternative, cost-effective models of infrastructure provision, at sufficient scale, are imperative to meet this goal.

Affordable energy, clean drinking water, dependable sanitation, effective public transportation, and reliable food systems will undoubtedly boost quality of life and enhance economic competitiveness throughout the developing world. But developing nations struggle to afford the capital and operational costs of sector-specific systems based on Western industrialized models. Moreover, many such systems incorporate fuels and technologies that are increasingly being implicated in climate instability and environmental degradation. Thus, even if they were affordable, these systems would be undesirable; with service lives extending anywhere from 50 to 150 years, they would lock in harmful emissions and wasteful practices throughout their useful lives.

But perhaps the prospect of undertaking infrastructure development with a clean slate offers a unique and timely opportunity to get things right. Many developing nations, particularly the least developed, are not yet encumbered by the material-, chemical-, and carbon-intensive legacy systems of industrialized society. This creates the potential to "leapfrog"—to bypass the shortcomings of past technologies and to move toward integrated, ecologically sound, socially responsive, and economical infrastructure systems.

Learning from Caracol, Haiti

Built with $300 million distributed through foreign-aid channels, including U.S. government contributions, and loans from the Inter-American Development Bank (IDB),[3] Haiti's Caracol Industrial Park was intended to be a showcase for the then relatively new Haitian government. Situated on a beautiful greenfield site in Haiti's Northeast Department, the public-private enterprise opened in 2012, after a year of fast-track construction.[4] In 2011, as he laid the cornerstone, President Michel Martelly announced, "Haiti is open for business."[5]

The U.S. government donated a 10 MW power plant to supply electricity for the campus,[6] and the plan was that the power, transport, water, and sanitation provided for the park would eventually be extended to neighboring villages, most of which lacked basic services. Photos from the project depict newly paved roads and a clean, brightly lit factory, where workers are assembling garments for export. To Haiti's citizens—eking out a living in a crippled, post-earthquake economy—as well as to the outside world, Caracol seemed like a godsend.

The site, which is bisected by the Trou du Nord River, had initially been recommended by an American consulting team and was later selected by the government of Haiti. Situated in the "breadbasket" of northeastern Haiti, the area has an abundant subterranean water supply, and consultants deemed it "capable of absorbing a large volume of treated water."[7] Subsequent studies, however, revealed that the park would have significant enough social and environmental impacts to persuade the IDB to change its risk rating from "medium" to "significant."[8] The work proceeded nevertheless—paving over a rural, 250 ha (608 acre) site where 386 families had once tilled some 300 plots of rich black soil, relocating the displaced households, and providing inadequate compensation for their loss.

According to a 2013 IDB report, the park did indeed have the "potential to cause significant and potentially irreversible negative environmental and social impacts,"[9] most of which were interrelated. First, the site's groundwater was being extracted from a widely shared aquifer—a withdrawal that could undermine the river's flow.[10] Second, effluents from a temporary wastewater-treatment plant could damage water quality, spoiling crops.[11] Third, the potentially compromised river emptied into the exquisite Caracol Bay, a critical natural habitat that hosts Haiti's last remaining mangrove forests, seagrass eel (a threatened species), and valuable coral ecosystems.[12] Finally, pollution from the park could decimate the fish nurseries on which local residents depend to supplement their subsistence-level farming.[13]

As of April 2015, only 6,200 of the anticipated 65,000 jobs projected for 2020 by the U.S. State Department had materialized.[14] But already, according to nearby communities that noted a marked decline in their water quality since the park opened, the project was placing undue pressure on the area's environmental resources.[15] Other risks—arising from a failure to understand the overall ecological context, as well as the interrelationships between the different components of the park—were of equal concern. The site was within a known earthquake fault zone and flood hazard area. Factory workers labored downwind of the power plant, which burned high-sulfur fuel oil,[16] emitting sulfur dioxide, nitrous oxide, and significant levels of particulate matter. This choice of technology was somewhat paradoxical, given that the coast surrounding the bay has "excellent potential for wind energy generation (average wind speeds of 7 to 7.5 meters/second) and optimal sunshine." Both wind and solar, however, were dismissed early on because the associated feasibility studies required longer time frames.[17] Lastly, social infrastructure—housing and transportation related to the park—remains inadequate. Whereas factory workers had initially been promised nearby housing, most of this has yet to materialize. Completed worker housing has since been found to be substandard.[18] In contrast to the foreign employees, who enjoy accommodations near the facility, Haitian workers continue to endure long commutes (see figure 1.1).

How did well-meaning participants plan and implement such a poorly sited, maladapted project, displacing hundreds and threatening one of Haiti's most valued natural and tourist sites? Through fragmented, uncoordinated processes, undertaken by multiple players across multiple sectors, in ways that failed to recognize critical interdependencies.

Nor is Caracol unique. Across the world, water, energy, sanitation, and transportation systems are "departmentalized": planned, designed, developed, and implemented by specialists who have little or no contact with their counterparts in other agencies or fields. Particularly in the Global North, such highly fragmented sociotechnical regimes have yielded infrastructure assets that may have satisfactorily met most needs for a time but whose inefficiency, carbon intensity, and lack of resilience will render them unsustainable in the long term.

But there is another way.

Figure 1.1
Workers awaiting their commute home at workday's end outside a Sae-A Trading Company factory building at the Caracol Industrial Park, Caracol, Haiti. Photo Credit: © Allison Shelley.

The Promise of Infrastructural Ecology

Ancient civilizations grasped the complex reciprocities of natural systems and created infrastructure that was modeled on, and worked in tandem with, those reciprocities. For developed and developing nations alike, this legacy of holistic thinking merits reconsideration and revival.

Over the last two or three decades, planners, engineers, architects, public officials, and others have been rethinking the fragmented, single-sector approach to infrastructure planning, design, construction, and implementation. Spurred by recognition of the harms associated with conventional industrialized infrastructure models, these actors are seeking alternatives that are resilient, sustainable, and economical—and that allow for increased community participation and more equitable distribution of development benefits. Infrastructural ecologies—integrated, holistic approaches to the provision of energy, water, sanitation, and transportation—are just such an alternative.

The vision of infrastructural ecology is based on the simple insight that the critical systems that serve settlements are analogous to the ecological systems that support nature's flora and fauna. As defined here, infrastructural ecology has its roots in industrial ecology: the study of energy and material flows in industrial systems. Taking as its model the circular flows of ecosystems, industrial ecology applies them to industrial processes in order to identify opportunities to reduce the consumption of energy and resources while minimizing pollution and waste. In a parallel manner, infrastructural ecology fosters exchanges across multiple sectors, enabling output from one sector to be used as input by another. The resulting "closed-loop" systems yield infrastructure that is more resilient and sustainable while offering higher economic, social, and environmental returns.

This book explores the potential of infrastructural ecology as a tool to address one of the most pressing needs of our time: the provision of basic services—including energy, clean water, sanitation, and transportation—in developing nations. Taking as our starting point the principles first articulated in Hillary Brown's 2014 publication *Next Generation Infrastructure: Principles for Post-Industrial Public Works*, we consider both the challenges and the opportunities of implementing infrastructural ecologies in emerging economies, highlighting successful examples that are already in place throughout the world.[19]

The Roots of the Infrastructure Gap

Even in resource-rich nations, meeting infrastructural needs continues to be a daunting task. The problems are almost unimaginably more severe in developing nations, which face a host of additional difficulties—rapid urbanization; lack of financial, organizational, and human capital; and diminishing natural resources. Emerging economies also share one difficulty with their developed counterparts: infrastructural silos.

Rapid Urbanization

In the megacities of the developing world, traffic chokes economic efficiency and dirty vehicles pollute the air, creating the familiar haze that was once common in all large cities but that has now been nearly eradicated in developed countries. Power is intermittent, often failing at times of greatest need; blackouts are common in the early evening, when homes and businesses are still fully active. Water service may slow to a drip on the hottest days of the year.

Nevertheless, the pull of cities has never been stronger. In the developing world, migration to cities stems from many causes—to improve economic circumstances, to participate in the global economy, or to avoid civil strife—but all center on the basic human

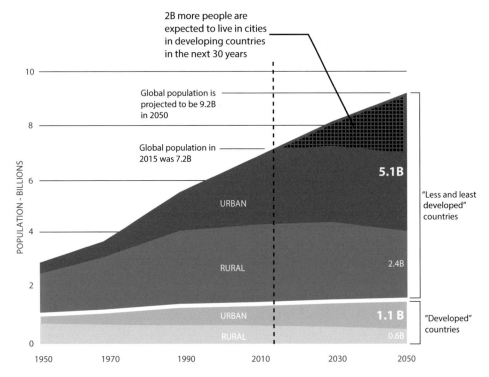

Figure 1.2
United Nations, Department of Economic and Social Affairs/Population Division, *Global Popula-tion Projections, 1950–2050. World Urbanization Prospects, the 2014 Revision, Highlights.* © 2014 United Nations, reprinted with the permission of the United Nations.

drive to increase well-being. In the face of this rapid population influx, however, many cities are struggling to provide basic infrastructure for their new urban dwellers. In many emerging economies, urbanization has vastly outpaced infrastructure investment. As a result, the urban poor often live in unsafe, unsanitary, and—too often—flood-prone conditions, while paying a large percentage of their income for services such as public transportation and water.

According to the United Nations' *World Economic and Social Survey 2013*, by 2050, more than 6.25 billion people will be living in cities (figure 1.2). And in the half century between 2000 and 2050, developing countries could add 3.2 billion urban residents—a number that exceeds the global population in 1950.[20] In 2005, the percentage of the world's population living in cities exceeded 50 percent for the first time in history; by 2050, that number is expected to increase to 66 percent.[21]

Although in some respects the current shift to cities is not unlike the urbanization trend that occurred in industrializing nations in the early 1900s, there are a number of significant differences. First, roughly 90 percent of the urban growth expected to occur by 2050 will be concentrated in Africa and Asia, with just three countries—China, India, and Nigeria—accounting for more than one-third of the total. Second, in many countries, the rate of change is up to three times faster than it was in the early twentieth century.[22] Finally, urbanization is occurring in areas that have fewer resources and slower economic growth than industrializing nations did a hundred years ago. For example, Burundi, a country that experienced a 2.5 percent increase in urbanization between 2010 and 2015, one of the highest annual rates of change in the world, is also one of the world's poorest countries.[23]

Much of this growth is expected to be unplanned, meaning that inhabitants will self-organize in informal settlements that lack adequate housing, basic infrastructure, and proximity to employment centers. Already, nearly one billion people in Africa, Asia, and Latin America live in such settlements,[24] where many dwellings lack reliable drinking water and sanitary sewer connections, trash may be strewn in the streets or at the nearest dumping ground, and narrow, winding dirt roads prevent entry of ambulances or other emergency vehicles. Retrofitting utilities in such settlements entails much greater disruption and higher costs—a problem that is exacerbated in emerging economies, where urbanization is often concentrated in the principal cities rather than being distributed across numerous smaller urban centers.

A small number of cities have managed to evade or overcome these urban infrastructural maladies; examples include several newly developed towns in China, as well as Kigali, Rwanda, and Medellin, Colombia, both of which have been fortunate enough to receive intense investment in rebuilding or recovery. Such cities, albeit few in number, offer powerful examples of integrated planning.

Indeed, despite its attendant difficulties, urbanization can offer an opportunity to provide the kinds of public services that will be the most effective engines of economic development—and to do so more efficiently.[25] Given increasing pressures to provide housing, employment, and basic services for those migrating to cities, governments are slowly recognizing that embracing sustainability, increasing resource effectiveness, and "internalizing externalities"—that is, accounting for the true monetary costs of development choices—are the only alternatives to continued forfeiture of the remaining stock of natural capital. Therefore, the challenge is to strategically plan for, finance, and implement the requisite infrastructure to achieve social and economic goals.

Lack of Fiscal, Organizational, and Human Capital

Most developing nations are caught in a "poverty trap," which prevents them from having access to their own infrastructure investment capital or that of others. First, their governments simply cannot tax an impoverished citizenry. Second, they may already be burdened by significant debt from prior administrations (the so-called debt overhang), which limits their creditworthiness. Finally, many are hampered by a legacy of unsuccessful structural adjustment policies and their accompanying austerity measures. Since the requisite investments outstrip the financing capacity of most developing nations, external support—in the form of aid, private investment, or both—is needed.[26] Yet many of these nations lack the organizational and human capacity to attract such investment.

In some cases, the governments of developing nations have failed to produce a secure environment that would be conducive to significant private or civil society participation in infrastructure development.[27] (Haiti serves as one example.) Lack of transparency and accountability may extend from public works ministers to meter readers. Funds may be decimated by bribery, embezzlement, and other forms of corruption, which are far too common in large civil works contracts. The risk of misappropriation of funds reduces the government's credibility in the eyes of foreign investors.[28]

Yet another major barrier is the limited professional class and the lack of educational programs to develop technical and managerial expertise to plan, design, and construct infrastructure projects. In some cases, an equally important impediment is the shortage of local workers capable of operating and maintaining the infrastructure. Thus, the implementation of infrastructure tends to be delivered by public sector agencies with staff that are less likely to have had the formal training to manage economic and financial risk or to shepherd projects through the complex institutional arrangements required to implement infrastructural ecologies.

Diminishing Natural Resources

Beyond the lack of financial, organizational, and technical capacity are two further obstacles: (1) the ongoing forfeiture of natural capital, a once-abundant patrimony, and (2) disproportionate exposure to the environmental and meteorological shocks attributable to climate instability. Both are undermining the ecosystem services on which infrastructural systems must draw.

The story of natural resource depletion in developing nations is a complex one, in which the various legacies of Western imperialism play a prominent role. One such legacy is a reliance on exports as the main source of income. These include agricultural products, such as sugar, coffee, and cotton, as well as primary commodities, such as timber and copper. Intensive ranching, overexploitation of fisheries, and destructive agricultural practices have decimated primal forests, threatened vital marine biodiversity, and depleted soils. Commercial logging and mining have not only stripped nations of their resources but have polluted their land and water. Tropical forests, virtual treasuries of potential wealth, are being effectively enclosed so that their biological material can be appropriated by pharmaceutical corporations, which then patent their genes for medicinal or cosmetic purposes.[29] Most tragically, emerging economies that export the bulk of their primary resources miss out on the economic rewards of value-added processing.[30]

Meanwhile, under the demands of structural adjustment—to which many developing nations are subject in order to restore their creditworthiness—emerging economies are under strong pressure to reduce fiscal deficits by exporting their natural capital, thus further degrading the commons.[31] Finally, in many cases, landless populations must draw on the only resources accessible to them, which are within the commons, and this also accelerates the extractive cycle.[32]

With regard to infrastructure, extraction of raw materials places extreme pressure on the very ecosystem services that infrastructure seeks to reinforce—and that would otherwise provide clean air and water, drainage, and even sewage treatment.[33] Depletion of natural resources also exacerbates the need for and the cost of basic infrastructure. For example, drinking water sources that have been polluted by mining require greater levels of treatment, which, in turn, calls for higher levels of technical expertise, more advanced technologies, and greater operational costs.

Reliance on resource extraction for income generation also leaves many communities vulnerable to conflict, which can have far-reaching consequences, particularly for infrastructure. A 2003 World Bank report noted that "when a conflict erupts, it not only sweeps away decades of painstaking development efforts but also creates costs and consequences that live on for decades."[34] In Liberia, for example, postwar urbanization increased the population of Monrovia, the capital, from 400,000 to 1.2 million. But the city's water treatment plant, severely damaged during the war and functioning at half its original capacity, could not begin to meet the new demand. Particularly at the confluence of conflict and urbanization, there is a profound need for a new paradigm to provide safe, reliable, and resilient infrastructure.

Infrastructural Silos

Since the advent of the industrial era, the dominant way of thinking has involved breaking down complex problems into component parts and placing each part in a box to await solutions implemented by specialists. Seeing each component as a discrete object, instead of as part of a unified system, has two effects: it obscures the primacy of interrelationships in complex systems, and it precludes our gaining a grasp of dynamic, complex system behavior—the acceleration of climate change being the prime example.[35] In a corollary effect, compartmentalizing forfeits efficiencies that could otherwise be realized through systems thinking.[36]

In the realm of infrastructure, the failure to engage in cross-sector infrastructural development is arguably a lost opportunity. As growing populations converge in urbanized areas, teeming, hastily built settlements will have ever greater need for interrelated, holistic infrastructure based on systems thinking. The failure to imagine context-specific, integrated systems will also hinder those left behind in underserved rural hinterlands.

Nor are industrial ways of thinking limited to those who design and build infrastructure. Organizations, for example, tend to undertake problems at scales they can manage and at which they believe they can have an impact. This pattern has particular relevance to developing countries, where infrastructure investment often depends on outside sources. Charities are often formed around a single issue and tend to fund stand-alone projects to meet single needs. Such projects—a well for drinking water or public toilets—often attract the most funding and provide tangible, if not necessarily far-reaching, results.

For many organizations, the metrics of success—the number of people served or the number of cases of a particular disease averted—are crucial to fund-raising. Dependence on such limited metrics, however, tends to produce small-scale undertakings rather than integrated approaches to service delivery. Such one-off projects forfeit valuable opportunities to strengthen outcomes through interconnections—and call out for the implementation of a new paradigm.

Despite some movement in the direction of more integrative approaches, host governments, investors, and multilateral banks and development agencies are still tied to old ways of thinking. In emerging economies, infrastructure is funded, built, and operated by a large group of discrete players, including, but not limited to, multilateral banks and private banks, philanthropies and other nonprofits, donor governments, host-country governments (at all levels), private businesses, and even individuals. The sheer number of actors—each with their own particular interests and agendas—can render integrated planning and design difficult. Moreover, infrastructure funding streams are typically allocated by sector, which eliminates the opportunity to capture potential synergies. It is not unusual, for example, for an infrastructure project to be conceptualized and designed in a foreign office by an organization that specializes in delivering a single type of infrastructure.

Because they tend to finance large-scale power generation, transportation, and water management projects (for both power and irrigation), banks, in particular, have profoundly shaped infrastructural models.[37] The selection of projects—and the technologies used in them—may also be affected by political pressure from donor and recipient countries alike. It is in some sense in the interests of developed nations, for example, to export their own industrialized models, despite the inevitable requirement for adaptation to climatic conditions as well as to social and cultural norms. Moreover, the infrastructure projects that are often selected for funding—such as roads, power plants, and water treatment plants—tend to be sector specific, and the interdependencies and relationships between such systems are rarely considered.

In sum, overcoming an entrenched, fragmented, and uncoordinated approach to planning, financing, and delivering critical services is a challenge in any environment but is particularly difficult for developing nations. Particularly in light of emerging economies' scarcity of economic, technical, and environmental resources, the call for the collective, contextual development of infrastructural services is all the more urgent.

Infrastructural Ecology: Why and How

The prodigious growth and prosperity of the Global North has come at the cost of the decline of the planet's natural capital; the pollution of its soil, waters, and atmosphere; and the potentially irrevocable alteration of its climate. Rising consumption of finite resources—by developed and emerging economies alike—leaves few alternatives but to determine how to decouple resource consumption from economic growth. Only then will it be possible to diminish the profound disparities across and within nations while simultaneously averting a legacy of even greater risks for future generations.

In other words, we must find a way to improve the quality of life in developing nations while curtailing both the local and global impacts of those improvements. In the face of potentially catastrophic climate change, developed and developing nations alike will need to learn how to live off the "interest" of the remaining stocks of natural capital; the depletion of finite resources is simply not an option. And because developing nations will host most of the population growth—as well as urbanization—in the coming years, they must be the locus of innovation and change.

There is a persistent misconception that economic and environmental values are at odds with each other—that economic growth must necessarily come at the expense of environmental objectives. But infrastructural ecology provides a means for addressing both without harming either. By capitalizing on synergies from interconnected flows of energy, water, and matter, multifunctional or networked structures offer a host of advantages:

• Sharing structural or mechanical components allows economies of scale, which saves on capital costs, increases land productivity, and conserves resources.
• Multifunctional structures save open space and reduce the sprawl associated with single-function development.

• Networked systems that rely on a combined workforce yield savings in operation and maintenance.

• Multifunctional and interconnected systems have built-in resilience to disruptions and system stresses, reducing the likelihood of failures and service outages.

• Multifunctional complexes (multiplexes) lessen the disruption, noise, and pollution associated with construction.

• Multiplexes offer opportunities for the integration of other civic amenities, including cultural, educational, and recreational uses.[38]

These advantages are so compelling that, despite significant obstacles, emerging economies throughout the world have put infrastructural ecologies in place. Typically, such projects are driven by a combination of economic, social, and environmental imperatives—as well as by the recognition that, in a resource-scarce environment, integrated, holistic infrastructure models based on nature will yield the most resilient, efficient, and sustainable systems.

Emerging economies vary widely in geography and climate, level of economic development, technical capacity, and historical context; nevertheless, infrastructural ecology projects developed in such contexts share a number of characteristics. First, because financial and environmental resources are limited, resourcefulness and invention are critical—a lesson that runs throughout the case studies in this volume. Second, given the high cost of imported fuel, materials, and food, closed-loop models—in which waste streams become inputs to secondary or tertiary processes—are fundamental economic drivers. Third, in the absence of external financial or technological support, community engagement often takes the form of complete self-reliance: local residents may create their own networked, ecologically based systems. And where governmental services are limited, community-based organizations may supplement them through informal local networks.

Industrial Symbiosis as a Model for Infrastructural Ecology: Two Examples

Industrial symbiosis, an outgrowth of industrial ecology, was developed to optimize processes across a group of industrial manufacturers; that is, to determine how efficiencies could be obtained through links between diverse producers. For example, funneling waste energy or material from one process for use in another reduces both input (of virgin material) and output (of emissions, waste, and pollution) from the whole system.[39] Through industrial symbiosis, industrial systems can, in effect, "evolve" to approximate the conservation behavior of natural ecosystems—in which nutrients (materials), water, energy, and by-products are circulated—to competitive advantage.

Discoveries at Kalundborg: Imperatives for the Collective Approach

First defined in 1989, industrial symbiosis is perhaps best exemplified by the ecoindustrial park in the seaside town of Kalundborg, Denmark. In the early 1960s, prompted by the potential for cost efficiencies, several of Kalundborg's industries began to share their residuals.[40] The 1,500 MW coal-fired power plant, which provided electricity and district heating to the town of 50,000, began to share its steam with an oil refinery and a pharmaceutical corporation. Meanwhile, sludge scrubbed from the power plant flues became the primary feedstock for a Sheetrock factory, replacing natural gypsum. Over the course of three decades, these initial partnerships grew to include more than 30 different exchanges of energy, water, and by-products between the colocated enterprises.[41] For example, a fish farm took advantage of the waste heat from the power plant, and neighboring farms substituted treated pharmaceutical sludge for costly fertilizer.[42] In time, information sharing and cooperation across diverse entities dramatically improved efficiency at the scale of the whole complex.[43]

What is significant about Kalundborg is the spontaneous emergence and development, over time, of a self-organized and self-regulated process.[44] The systems-oriented framework did not originate all at once from a centralized vision. Instead, it emerged gradually from disparate entities that shared a collective motive: producer profit. The potential for higher yields gave rise to a peer-to-peer network of trust and social cohesion. According to one report, "Personal and long-term professional relationships [were] necessary to maximize synergies."[45] The opportunity factor—the relatively short distances separating the businesses, and the associated efficiencies in transport and exchange—was also critical.

Industrial ecology offers several ways to reduce greenhouse gas emissions: through direct energy savings from colocated companies sharing a single source of combined heat and power; through heat recovery from steam, condensate, and other sources; and through indirect energy savings accrued from reduced transport and localized materials recycling. Assessments of Kalundborg have shown, over time, that participating manufacturers and utilities have much lower environmental impacts than nonnetworked entities, thanks to the conservation of materials and avoiding the pollution of air, water, and soil. Among the annual outcomes are the following:

- Reducing carbon dioxide emissions by 275,000 metric tons (303,135 tons)
- Saving 3 million m³ (792.5 million gal) of water (through recycling and reuse)

• Recycling 150,000 metric tons (165,346 tons) of gypsum (from desulfurizing the power plant flue gas)
• Producing 200 metric tons (220.4 tons) of fish
• Replacing 70 percent of the soy protein used in the feed mix for 800,000 pigs (as well as other animals) with yeast slurry from the pharmaceutical plant[46]

The aggregate value of the savings is estimated at $15 million annually and at $310 million cumulatively since the program's inception.[47]

Hammarby Sjöstad: Implementing Infrastructural Ecology
In the years since the Kalundborg ecoindustrial park was founded, several industrializing nations (e.g., China, Brazil, India) have taken up the concept as a means of increasing the sustainability of typical industrial practices. Other urban-scale applications, however, demonstrate the potential of infrastructural symbiosis, and this section considers one of the most successful examples.

When the International Olympic Committee proposed a former industrial site on Stockholm Harbor as the site of the Olympic village for the 2004 games, the city of Stockholm answered the call, committing to building an "environmentally well-adapted city district . . . at the forefront of international strivings towards sustainable development in densely populated urban areas."[48] Despite losing its Olympic bid, the city nonetheless moved ahead aggressively to control the metabolic flows (energy, material, water, and waste) of the new mixed-use, compact development district, Hammarby Sjöstad. When completely built out, which is scheduled by 2017, it will accommodate 25,000 residents and 5,000 workplaces.

The city's vision for Hammarby Sjöstad was based on the goals articulated by local authorities in 1992 in accordance with Agenda 21, the United Nations' sustainability plan. It was also compatible with the Natural Step, a popular sustainability platform developed in Sweden in 1989 by scientist Karl-Henrik Robèrt, which defined sustainable limits derived from the laws of thermodynamics and the observation of natural cycles.[49]

With broad local political support, the city created environmental (including carbon-reduction) targets, chose infrastructure technologies (some of which were new, others adaptations of existing plants and distribution systems), and established a design team.[50] In keeping with the current administration's environmental sensibility, the team strove for integrated solutions. Team members included city planners, landowners, building developers, real estate administrators and operators, and representatives from the municipally owned entities that provided Stockholm's water, waste management, energy, and regional transport. The city specifically asked the municipal entities to consider how their infrastructure assets could work in tandem in order to reduce energy and

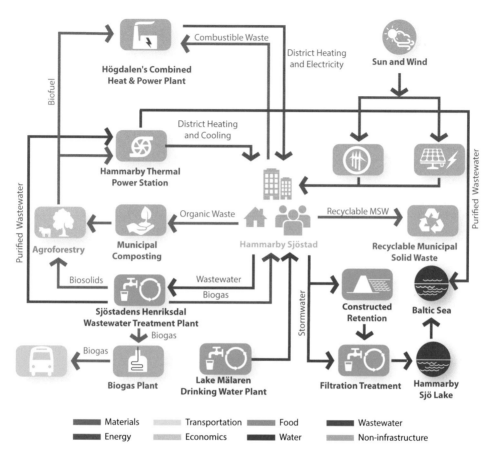

Figure 1.3
The "Hammarby Model," integrated urban systems for Hammarby Sjöstad, Sweden. Redrawn by
Logman Arja from the original by Lena Wettrén, Bumling AB.

environmental impacts. The collaborative development of solutions, through negotiation, made it possible to avert the typical fragmentation of sectors.[51]

This exemplary planning process, which was based on systems thinking, yielded a
symbiotic infrastructural solution now known as the Hammarby Model (figure 1.3).
Among the planning objectives were the following:

Natural cycles should be as closed at as local a level as possible; energy should be derived from
renewable sources, and as far as possible, obtained from local sources; and sewage is to be utilized
for energy extraction, and its nutrient content returned to agricultural soil.[52]

To an unprecedented extent, Hammarby Sjöstad was planned to integrate urban systems in such a way as to cycle flows of material, water, and energy internally. All elements in the systems—from heat and power production, to transport, to water supply and treatment, to waste management—were designed or adapted to work in tandem to reduce the overall district metabolism (see figure 1.3).[53]

Since 1970, Högdalen—a combined heat and power plant—has been incinerating much of Stockholm's domestic waste and providing Hammarby with electricity and district heating. There are also two supplementary heating sources: the heat that is extracted from the wastewater treated in the Henriksdal treatment plant and that from centralized oil-fired or electric boilers.[54] At Henriksdal, which continues to serve other parts of Stockholm as well, sludge is processed to yield biogas, which is used for local cooking or further refined to serve as biofuel for Stockholm buses.

Other synergies were part of the initial model but were eventually dropped. One idea would have used biosolids from the wastewater sludge as agricultural fertilizer, but instead they were diverted for use as filling material for mines. Another implementation shortfall was the failure to achieve scale in renewable energy technologies (photovoltaic panels, solar thermal panels, and wind turbines) because of excessive cost. The Hammarby Sjöstad system thus falls short of being self-sufficient in terms of locally generated energy.[55]

Although studies have shown that the district's closed-loop systems, which made use of the secondary energy resources in waste, significantly reduced metabolic flows,[56] Hammarby Sjöstad's planners ultimately failed to collect sufficient data to assess the system's overall performance—a step that would have contributed significantly to the establishment of future districts.[57] Nonetheless, the initiative offers a meaningful template for infrastructural ecology that can be revisited and strengthened by others.

The Organization of This Book: The Five Objectives of Infrastructural Ecology

To support broader adoption of ecologically sound, socially responsive, and economical infrastructure in developing nations, this book explores successful projects that have already been implemented. Most are contemporary, but each chapter also features exemplary preindustrial systems, highlighting the surprisingly long history of resilient, economical, sustainable, and community-driven infrastructure. In each case, the text highlights lessons learned and best practices, offering transformative models that have the potential to restore planetary well-being in the face of resource depletion, energy volatility, and declining availability of water.

The alternative vision for infrastructure development proposed in this volume calls for integrated, holistic solutions conceived in accordance with the following five objectives:

1. Relational solutions: hybridized, integrated, and reciprocal systems
2. Ecological alignments: systems modeled on and associated with natural processes
3. Low-carbon processes: systems with reduced reliance on nonrenewable fuel
4. Resilient constructions: systems prepared for climate instability
5. Codevelopment: inclusionary and participatory projects

Chapters 2 through 6 focus on each objective in turn, using case studies to illuminate how the objectives can be implemented. Although the cases in any given chapter may illustrate more than one aspect of infrastructural ecology, they were selected as particularly compelling examples of that chapter's main focus.

Chapter 2 delineates the crux of infrastructural ecology: the use of single solutions to solve multiple problems and thereby gain efficiencies and economies of scale. The examples are drawn from a range of eras and sectors—from a seventeenth-century bridge in Iran, to a solar array spanning a reservoir in India, to an ecoindustrial park in China. Chapter 3 focuses on holistic, "soft path" alternatives for water management, including capture and storage, green infrastructure, and water reuse and nutrient recovery. Together, such approaches can protect against flooding, improve water self-sufficiency, and enhance security of food and livelihoods. Chapter 4 focuses on "post-carbon" infrastructure for power, heat, and transportation. The approaches highlighted in the case studies include microhydro in rural Nepal, domestic solar hot water in Brazil, biogas derived from wastewater in Nigeria, and a solar rickshaw in India. Chapter 5 focuses on climate adaptation, a topic of acute importance in developing nations, many of which are located in the latitudes that are most at risk. Among the approaches to adaptation and mitigation featured in the chapter are multipurpose tunnels in Indonesia, a "floating city" in Nigeria, and wastewater reclamation in Jordan. Chapter 6 focuses on community-based approaches to infrastructure as a means of ensuring that projects meet the needs of those they are meant to serve. Full, meaningful participation is key to developing culturally appropriate solutions, achieving equitable distribution of benefits, and building local capacity. Among the projects highlighted in the chapter are a coastal resilience initiative in Viet Nam, a sustainably managed forest commons in Mexico, and a community-initiated water supply scheme in Tanzania.

Chapter 7 addresses the financial, institutional, and political barriers to implementing infrastructural ecology projects in emerging economies and offers guidance for overcoming such obstacles. Chapter 8 also addresses implementation, but from the perspective of policies and strategies; it provides specific recommendations, objective by objective, for creating the kinds of environments that will support infrastructural ecologies.

Throughout the book, diagrams illustrate the flows of energy, water, materials, food, and waste—demonstrating not only the functions and structure of each element but

also the ways in which subsystems can be integrated to yield higher-performing infrastructural ecologies.

Imperatives for Infrastructural Ecologies

It is difficult to fathom the speed and scale of current and projected urbanization. Given
UN projections of population growth—an additional 2.5 billion people by 2050—it
will be necessary to build a city capable of housing six million people once a month
for the next 35 years, along with the roads, energy and water supply, sanitation, and
other systems needed to support it.[58] Unless persuasive new development models take
hold, greenhouse gas emissions will grow exponentially, particularly in the developing
world. The combination of climate change and the loss of arable land to urban expansion will further stress the food and water supplies of populations that are already poorly
nourished.

   We view these pressing scenarios not as inevitable disasters but as potential drivers
of positive change. An integrated, holistic approach—defined here as infrastructural
ecology—holds the promise of economically coupling multiple functions; making greater
use of natural processes; following a low-carbon path; promoting long-term climate resilience; and incorporating participatory processes and local knowledge. The case studies
in this volume illustrate the potential of just such an alternative paradigm. By employ-
ing a macroscopic perspective that makes the most efficient use of limited resources,
infrastructural ecology answers an urgent call for networked, synergistic services that
will meet current and future needs.

Solving for Pattern: From Interconnected to Symbiotic Systems

A good solution solves more than one problem, and it does not make new problems.
—Wendell Berry, *The Gift of Good Land*

In a nation beset by poor road and rail connectivity, Bangladesh's Jamuna multipurpose bridge has forged a vital conduit between the eastern and western halves of the nation, which had long been divided by the Jamuna River. Until 1998, when the 4.8 km (2.98 mi) road and rail link was completed, crossing the Jamuna River entailed a ferry journey that lasted between 16 and 30 hours; the bridge reduced travel time to approximately 4 hours. Moreover, carbon emissions dropped—and, most importantly, the once-marginalized northwestern part of the country was joined to the eastern portion, promoting trade and investment, spurring job creation, and substantially lowering local poverty rates.[1]

Built with support from the World Bank, the Asian Development Bank, Japan's Overseas Economic Cooperation Fund (OECF), and the government of Bangladesh, the bridge offers two-lane carriageways, a dual-gauge railway (accommodating trains of two separate gauges), high-voltage and telecommunications cables, and a high-pressure natural gas pipeline (see figure 2.1).[2] The decision to combine the various functions was made after feasibility studies revealed that none of the individual projects were independently economically viable.[3] The bridge's successful combination of transport and utility functions is largely unique, not only in the developing world but in industrialized countries as well.

The phrase "solving for pattern," which was coined by Wendell Berry, refers to finding solutions to problems by understanding the patterns that connect them. Holistic thinking—that is, capitalizing on proximate relationships and building on

Figure 2.1
Jamuna multipurpose bridge, Bangladesh. Photo Credit: Nusrat Kabir, Wikimedia Commons.

interdependencies and reciprocities—epitomizes solving for pattern. As a framework for optimizing infrastructure investments, solving for pattern results in a solution that, instead of serving a single function, serves multiple ends. When compared with stand-alone, single-purpose systems, multipurpose systems—which may share space, structures, operating strategies, or inputs and outputs—deliver greater benefits at less cost.

The examples highlighted in this chapter exemplify infrastructure that is more resilient, cost-effective, and sustainable than conventional approaches. Drawing on systems thinking, they make more intensive use of limited space while offering greater social, environmental, and financial returns. This chapter examines a variety of these systems, from the simplest arrangements, in which unrelated entities achieve efficiencies or economies through colocation, to instances in which one service effectively "hosts" another, to symbiotic designs in which one entity uses the wastes of another.

Preindustrial Ingenuity: Multifunctional River Crossings and Agro-Infrastructure

Multifunctional civil undertakings can serve as models for contemporary infrastructure investments. Products of resourceful preindustrial societies, they resolve multiple problems with single solutions—accommodating different modes of use, spatially concentrating a variety of social and economic functions, and exploiting useful synergies. The early examples highlighted in this section take advantage of closely observed natural phenomena and are finely attuned to context—including local resources, hydrology, and climate. Like their more recent counterparts, which will be considered later in the chapter, they illustrate the often unrecognized potential of integrated thinking across infrastructural systems and sectors.

Isfahan's Khaju Bridge

Preindustrial river crossings—superstructures harboring shops, multistory housing, and civic spaces—supported the dense mix of uses that their nearby cities required. Some also incorporated mechanisms for flood control, irrigation, and energy production.[4] Compared with these hybrid feats of infrastructural engineering, today's single-function investments seem like missed opportunities.

Located along the Silk Road in Isfahan, Iran, the seventeenth-century Khaju Bridge (figure 2.2) features 23 supporting arches that span the Zayandeh River. Completed in 1687, on the foundations of an older crossing, the bridge combines transportation, hydraulic regulation, and civic structures, and may be one of the world's first multimodal double-decker bridges. The main aisle of the upper story was designed for animal-drawn vehicles; pedestrians used shaded outer arcades, which led to octagonal pavilions embellished by frescoes and tile work—princely stations that afforded upstream and downstream vistas for festive aquatic events. On the lower story, linked, shaded galleries—cooled by the nearby passage of the water—still feature shops, and serve as popular social spaces. On Fridays, the galleries are filled with the sounds of Isfahani men singing. Steps cascading along the downstream side provide seating, as well as public access to the river.

In arid Iran, water scarcity drove the development of sophisticated irrigation techniques: weirs, canals, reservoirs, and some 50,000 *qanats* (ancient tunnels, of which 30,000 are extant) channeled water from the high-elevation snowmelt.[5] Like other river crossings of its time, Khaju Bridge also functions as a dam. During the droughts of summer, the sluice gates are closed, and water impounded behind the upstream side irrigates nearby farmland and ornamental gardens. Deep channels between the arches

Figure 2.2
Khaju Bridge, downstream view. Photo Credit: Khosrow Bozorgi, 1999, courtesy of the Aga Khan Visual Archive, Aga Khan Documentation Center at MIT.

allow passage of river water, even during low flow. During high water, stepped spill-ways on the downstream side dissipate the river's energy, preventing scouring of the bridge's base while aerating the water, helping to purify it.[6] A multimodal bridge, a weir, and a heavily used gathering place, Khaju Bridge merges complementary func-tions into a splendid public monument that has yielded economic, social, and environ-mental dividends down through the centuries.[7]

The *Chinampas* of Mexico

Chinampas—the integrated waterways and agricultural plots of pre-Columbian Mesoamerica—offer further examples of compound utility. Other pre-Columbian civi-lizations also relied on these systems of raised fields: the Mayas used them in their seasonally inundated fertile lowlands to reclaim the waterlogged soils that would otherwise have been flooded for months at a time.[8] For centuries, *chinampas* also sus-tained the dense urban settlements of Tenochtitan, the island-city Aztec capital in Lake Texcoco,[9] as well as other lake-based civilizations in the basin of Mexico.

Versatile constructions that integrated hydraulic, transportation, and sanitation functions (dams, aqueducts, canals, and landfill), as well as advanced farming techniques (irrigation, fertilization, and aquaculture), the *chinampas* embodied a systems approach, permitting year-round food production and fostering the capacity to tolerate climate variability. They were still in place in 1519, when the Spaniards entered the basin of Mexico, and continue to exist in limited form today.

In Lake Texcoco, the Aztecs built rectangular *chinampas* plots—which ranged from about 9 to 15 m (30 to 50 ft) in width and as much as ten times that in length—by excavating the rich, shallow soil of the lake bed, which also deepened the interconnecting water channels, benefiting transportation. The excavated mud—mixed in alternating layers with decayed aquatic vegetation (grasses, water lilies, or bulrushes), rock, and human waste—was mounded into beds bounded by woven reeds that were held in place by wooden stakes. To prevent erosion, the plots may have been edged with willow trees, whose roots would have helped stabilize the plots a foot or so above the lake level.[10] The trees could also have served as windbreaks, increased biodiversity, and provided a ready source of firewood.[11]

As the *chinampas* grew outward into the lake from the center of Tenochtitlan, the reclaimed swampland—crisscrossed by canals—came to be the most densely inhabited part of the city. The canals served as public waterways, as well as the primary means of irrigation, and were controlled by a system of stone weirs and aqueducts that kept water levels stable.[12]

Across this vast, gridded network of raised plots, the Aztecs practiced a space-efficient and intensive form of agriculture. Seeds that had already germinated in floating nurseries were transplanted into the fields, an approach that allowed from two to seven crop yields a year of maize, tomatoes, peppers, manioc, edible greens, and root vegetables, at a scale capable of feeding large urban populations.[13] Because of their integrated construction, *chinampas* were, for the most part, both self-irrigating and self-fertilizing—characteristics that further supported year-round cropping. The proximity of canal water (see figure 2.3) reduced the need for labor-intensive irrigation. Instead, the plants obtained moisture through capillary action, drawing from the lateral seepage of the canal.[14]

Crops were fertilized by the original aquatic plant matter that had been mixed into the soil, as well as by human waste (and in some regions by bat guano from caves).[15] Reapplications of lake mud also supplied nutrients that had been deposited by aquatic life. Synergistic returns from this constructed ecosystem included increased nitrification, reduction of pests and root rot, moisture retention, easier harvesting and weeding, and fertility enhancement.[16] The routine dredging also kept the waterways free of

Figure 2.3
Chinampas and canals, 1912. Photo Credit: Karl Weule, © 1926 Karl Weule, Leitfaden der Voelkerkunde, Leipzig, 1912.

obstructions. Finally, the canals themselves provided the population with a protein diet of fish, turtles, crustaceans, and waterfowl.[17]

Significantly, the *chinampas*, by virtue of their layout, proportions, self-shading, and adjacency to water, created a favorable, self-sustaining microclimate that was cooler in summer. Moreover, they were warmer in winter, as the thermal storage of the water (daytime heat reradiated at night) ameliorated cold temperatures and reduced the risk of frost for the crops.[18] *Chinampas* may also have housed "live-work" populations: cultivators are thought to have had huts on the platforms as well.[19] In sum, this highly integrated, multifunctional agricultural landscape enabled a burgeoning urban population to overcome geophysical threats to its survival, including frost, variable rainfall, and soil depletion.

Despite the obvious advantages of the *chinampas*, not only did the majority of them disappear from the Americas, but also the practices associated with them—the skillful, labor-intensive manipulation of climate, microclimate, hydrology, soil, and living

matter, with its legendary levels of agricultural density, diversity, and productivity[20]—
were largely erased from history. Only in recent decades have we overcome our distaste
for swampy, marginal lands; instead of draining them, we now prize them for their
ecological richness and diversity.

Similar preindustrial approaches to raised-field cultivation are known to have existed
in Asia and Africa. For example, the dikes of China's Pearl River delta and the raised-
field agriculture practiced in what is now Cuvette Province in the Republic of the Congo
relied on the rich soil drained from marshlands.[21]

The agricultural benefits of such practices have long been apparent. Several of
the chapters that follow highlight the reemergence of what might be called an agro-
infrastructure: the colocation of agriculture with other infrastructural networks. Such
combinations of food production or biomass production with other functions, such as
energy production and water desalination, have significant implications for the future.

Simple Integration: Colocated Systems

In colocation, two or more processes or services that would otherwise be unlikely to
occur in the same space are associated for mutual advantage. Sharing real estate or construction elements can allow colocated projects to accrue savings—and may sometimes
render otherwise infeasible projects practical.

Multi-Utility Tunnels

A prime example of a basic colocated system is the "utilidor" or "multi-utility tunnel"
(MUT)—a below-grade network that houses cables and pipes for a range of functions,
including water distribution, telecommunications, and district heating and cooling.
Typically consisting of concrete or steel culverts of varying sizes, these flexible systems
avoid the redundant costs, health and safety concerns, and disruption associated with
conventional open-cut excavation and filling, and in emerging economies, they offer a
more economical means of keeping pace with escalating utility demands.

Early prototypes of MUTs were developed in nineteenth-century London and Paris.
These were followed, in the 1930s and 1940s, by more sophisticated systems in Madrid
and Moscow.[22] Singapore was the site of the first comprehensive, large-scale network
constructed in Southeast Asia (figure 2.4). Completed in 2005, the $1.26 billion project
consists of two tunnels, which are approximately 2.5 m (8.2 ft) below grade and have
a combined length of 3 km (1.86 mi). The tunnels, which serve the Marina Bay district, combine power lines, water and wastewater distribution, pneumatic refuse collection, and telecom cables. The project also incorporates cold water for district cooling,

Figure 2.4
Marina Bay common service tunnel utility services, perspective. © Urban Redevelopment Authority, Singapore. All rights reserved.

thereby freeing up space on the roofs of buildings—otherwise used for cooling towers—for more creative purposes. In Singapore, the goals of optimizing land use (by freeing up road verges that would otherwise have been set aside for utilities) and expanding the use of underground space made the solution virtually inevitable.[23]

In renovating Connaught Place, New Delhi's high-end commercial center, completed in 2014, local officials upgraded urban supply systems with a 1.2 km (~.75 mi) MUT. Seven meters (22.9 ft) in diameter, the $85.7 million tunnel features neatly organized stacks of color-coded utilities on either side of a 1 m (3.28 ft) access and maintenance corridor. It accommodates electric, gas, and water pipelines (green for irrigation

and recycled water, blue for potable water), along with telecom cables, and has designated ducts for firefighting. Spaces have been constructed at one-block intervals for the
eventual placement of transformers, and provision has been made to support central
cooling in the future. Sewerage and storm drainage pipes run parallel to but outside
the MUT;[24] however, monsoon rains have made it necessary to retrofit the tunnel with
sump pumps.

Twenty-five kilometers (15.53 mi) south of the Malaysian capital of Kuala Lumpur
lies Putrajaya, a planned city that is now the nation's administrative center. With its
state-of-the-art fiber-optic network, this "intelligent" city supports high-capacity digital information and communications technology (ICT) throughout all government
facilities, as well as local public wireless access.

Among the city's signature projects is an MUT undertaken through a joint venture
on the part of five local utility companies that share capital and maintenance expenses.
Spanning 11.4 km (7.1 mi) and housing power, telecommunications, gas, water, and
chilled water pipes for district cooling, the MUT also features a sophisticated security
system that allows continuous monitoring of conditions within the tunnel.[25]

The MUT is hardly a futuristic idea—the concept is neither new nor complex. Yet it
remains underused in industrialized nations, largely because of the high cost of rerouting
existing utilities. Developing nations, however, may have the opportunity to avoid the
cut-and-fill approach and instead bundle various services—including electricity, telecommunications, and water—in underground trenches situated within new right-of-
way constructions, whether for road or rail. (Telecommunications can also be integrated
with other systems through aerial facilities—towers, pylons, and power lines.)[26]

Although higher initial costs and the difficulties of cross-sector coordination pose
challenges,[27] where public-private partnerships are feasible—and where utilities can be
planned and managed centrally—MUTs offer long-term economic advantages, including long service life (more than 100 years); economies of scale; and ease of maintenance, expansion, renewal, leak detection, and control. Moreover, when MUTs are built
in tandem with new roads, their construction is cheaper and faster.[28] As long as utility
density is high and street openings for utility connections are frequent, the combination of economic benefits with reductions in environmental damage, social and economic disruption, and safety concerns can tip the balance in favor of MUTs.[29]

A "Smart" Project in Kuala Lumpur

The SMART (stormwater management and road tunnel) system completed in 2007 in
Kuala Lumpur, a densely populated city of 1.5 million people, serves dual functions.
Most of the time, the 13.2 m (43 ft) diameter tunnel—which features a double-decker

motorway in its middle third—is used to divert vehicular traffic from the highly congested central business district. With a daily capacity of some 30,000 vehicles, the tunnel cuts travel time by as much as one-third while reducing the pollution associated with cars idling in traffic jams.

During seasonal downpours, however, the tunnel is used to eliminate stormwater from the highly flood-prone downtown. It has a storage capacity of 3 million m³ (3.92 million yd³) and is capable of diverting 90 percent of urban stormwater during major flooding. A sophisticated control system allows rapid switching of operating conditions to satisfy both modes of use.

When SMART is in standard mode, with little or no rainfall, no water flows into the tunnel. In the event of a moderate storm, rain gauges in the floodwater catchment zone trigger the diversion of water from the city's holding ponds into the tunnel. In this

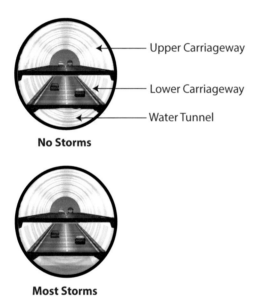

No Storms

Most Storms

Major Storms

Figure 2.5
Operation of the SMART Tunnel, Kuala Lumpur. Redrawn by Logman Arja, after an image from Storm Water Solutions, http://www.estormwater.com/kuala-lumpur's-smart -tunnel-expected-be-operational-middle-next -year.

second mode, excess stormwater enters the lowest section of the tunnel while motor traffic continues to flow above it. The system relies on real-time information from 28 sites that provide data on rainfall and river flow, including early warning of possible flooding.[30] During monsoons, traffic is evacuated, and the entire tunnel fills with stormwater (figure 2.5). Within 48 hours, most of the waters recede through the force of gravity; pumps are used to draw out the rest. The tunnel is then desilted, washed down, and readied for vehicular traffic.[31]

At 9.7 km (6 mi), the SMART is currently the longest multipurpose tunnel in the world. A joint venture between the federal Department of Irrigation and Drainage and the Malaysian Highway Authority, the tunnel cost $515 million; estimated savings over a 30 year period, however, have been valued at $1.58 billion for flood damage prevention and as much as $1.26 billion in productivity gains (realized through decreased traffic congestion).[32]

It was the motorway that made the diversion project economically feasible: a single, dual-use project meant less disruption and lower costs. Moreover, since much of the construction costs are being recovered through tolls, the vehicular use pays its own way.[33] In 2011, UN Habitat, the United Nations Human Settlement Program, added the SMART project to its Scroll of Honor for its highest and best use of resources.

As testament to the tunnel's success, the administration of Jakarta City is considering constructing two underground dual-purpose tunnels—11 m (36.08 ft) in diameter and 12 km (7.45 mi) in length, costing Rp 24 trillion ($2.92 billion)—to manage overflow from the Pesanggrahan and Ciliwung Rivers.[34] In highly urbanized areas, the economic and environmental cases for double-duty infrastructure projects like the SMART are potentially strong but have rarely been explored.

Commensalist Associations

In nature, one organism sometimes benefits from joining with another in ways that do not compromise either; the term for this is *commensalism*. Similarly, in the realm of infrastructure, one system can "host" another through colocation. The result—more rapid and cost-effective rollouts of critical services—offers yet another example of the benefits of solving for pattern.

ColaLife: In Zambia, Medicine Hitches the Last Mile

In low-income countries, the rural segment of what economist Paul Collier refers to as the "bottom billion" is often disconnected from the formal networks that deliver public and commercial services. When it comes to transportation, for example, those who live at "the last mile" must often rely on informal, sometimes unreliable arrangements with vehicle owner-operators—trucking traders and entrepreneurs who dominate the transport industry. Health care, which depends on the transportation of medicine, equipment, and fuel, often suffers as a consequence.

In rural Zambia, where diarrhea is the second-leading cause of child mortality (15 percent of deaths of those under five), the independent nonprofit organization ColaLife has stepped into the breach. In an effort to reduce the under-five mortality rate and to improve mothers' health and awareness of health risks, the organization's founder, Simon Berry, found a way to deliver crucial medicine to remote communities. His idea—sparked by the realization that Coca-Cola, the universally appealing soft drink, reaches even the farthest corners of the globe—solves for pattern by capitalizing on transportation mechanisms that are already in place.

ColaLife partners with last-mile distributors whose buses, trucks, carts, and even bicycles deliver crates of the soft drink to small retailers in isolated areas, piggybacking life-saving medicines on Coca-Cola deliveries.[35] Wedge-shaped sleeves were customized to fill the spaces between the bottle necks in soft drink crates, adding no extra volume and insignificant weight (figure 2.6). Designed for use by mothers and aid workers, the sleeves (known as AidPods) contain oral rehydration salts, zinc supplements, soap, and an instruction leaflet, and are reusable and recyclable. Thanks to the support of the regional Coca-Cola bottling company, which shares the modest record of corporate social responsibility associated with its parent organization, a public health product gets an easy lift along a private sector supply chain.

ColaLife's initial, 12 month trial was successful—26,000 kits were sold in two remote areas—and the effort is currently being scaled up, with a goal of reaching 22 more districts.[36] As in the case of soft drink distribution itself, small profits accrue at each stage to those who pass along the AidPod, generating a secondary value chain. By using design to optimize the use of crate space, ColaLife offers an open-source model for the delivery of other types of goods to underserved populations. Taking advantage of an existing distribution system by combining unrelated goods is an admirable example of the "sharing economy."

M-Pesa: Banking on Help from the ICT Sector

In emerging economies, lack of access to and participation in local, national, and global financial systems is a well-known barrier to economic growth, income equality, job creation, and human development. Those who are cut off from markets by physical distance—because conventional banks have been unable (or unwilling) to serve remote or sparsely populated areas—and those who have irregular income streams

Figure 2.6
ColaLife's AidPod shown in Coca-Cola crate. Photo Credit: Courtesy Simon Berry, ColaLife.

have generally been deprived of opportunities to obtain credit and sell products or services.

According to the World Bank, as of 2012, at least 2.5 billion people worldwide lacked access to basic financial services.[37] Christine Lagarde, managing director of the International Monetary Fund (IMF), claims that this number translates to more than one-third of the global population, a large percentage of whom are urban dwellers.[38] The World Bank has noted, for example, that the majority of Latin America's urban population is "unbanked."[39]

In the heart of Africa, researchers at two telecommunications companies had noticed that individuals in several central African nations had improvised ways to use mobile airtime to transmit money simply by transferring mobile minutes to friends or relatives, who then used or resold the time. In 2007, emboldened by this model, the mobile network operator Vodafone proposed a cross-sector innovation known as M-Pesa, which is Swahili for "money." Piggybacking on the speed, security, and transparency of the mobile sector, Vodaphone extended its digital platform to support microfinance, allowing it to be used for bill payment, receipt of wages, currency exchanges, and other money transfers. The cost-efficient program scaled rapidly.[40] By 2014, for example, about three-quarters of adult Kenyans were connected to banking services via mobile networks,[41] and the 18.2 million M-Pesa clients in Kenya alone had yielded $143 million in revenues.[42]

In 2012, Safaricom, the largest mobile network operator in Kenya, and the Commercial Bank of Africa developed a partnership, M-Shwari (Kiswahili for "calm"), through which M-Pesa customers can establish interest-bearing savings accounts and even obtain short-term credit. The program also enables family members working abroad to send money home. Such technological cross-pollination sets the stage for future innovations across diverse sectors.

According to one report, mobile money accounts have widely outpaced traditional bank accounts in Africa. Meanwhile, mobile money services have been rolled out in nine other national markets—from Bolivia to Brazil, Egypt, Ethiopia, Tajikistan, and Togo.[43] And in a novel reversal, digital currency is being introduced in industrialized nations. Vodafone expects that an e-money license will allow it to operate similar financial services in Central and Eastern Europe, with Romania—a nation where an estimated seven million people still rely on cash—as the first trial.[44]

Water Pumping as Child's Play in Colombia

Beginning in the early 1970s, the new village of Las Gaviotas, which is situated 500 km (310.69 mi) east of Bogotá, in *los llanos*, Colombia's inhospitable savannah, responded to the Organization of Petroleum Exporting Countries (OPEC) oil embargo by developing a range of off-the-grid inventions. With early support from the United Nations Development Programme, the village's founder, Paulo Lugari, enlisted teams of engineers, artists, students, indigenous peoples, and orphans from the streets of Bogotá to develop innovative "appropriate technologies" that would provide essential infrastructural services for this groundbreaking settlement.[45]

Among the greatest challenges in the region was the collection and transport of water, a task that had traditionally fallen to women and children. The resourceful collective came up with a novel solution: a pump that was capable of extracting water from aquifers that were six times deeper than those that could be accessed by traditional means—and with less effort. Moreover, during the dry season, the pump provided more affordable and sanitary access to groundwater.

Figure 2.7
Visitors operate seesaw pump in the village of Gaviotas, Colombia. Photo Credit: © John Winters.

In the conventional pumps used in the region, a heavy piston moved within a pipe, whereas in the Gaviotas pump, the piston remained in place while a lightweight and inexpensive PVC sleeve moved around the piston. And atop the pump, the designers positioned a lightweight beam fashioned in the form of a seesaw (figure 2.7). By playing on the seesaw, the local children pumped water into an adjacent community cistern, recapitulating their traditional role as water collectors but with greater joy and considerably less effort.

These "parafunctional" objects—simple playthings that transmit lessons in sanitary hydrology—have since been replicated across Colombia.[46] By 2003, seesaw pumps were being used in more than 700 villages.[47] Today, seesaw pumps are manufactured in India for use in rural schoolyards. In Africa, however, the child-powered merry-go-round version has met with less success. Despite initially promising results, lack of user input and maintenance training led to the failure or rejection of the equipment.[48]

Les Anglais: An Opportunistic Power Arrangement in Haiti

As of 2014, about 75 percent of the Haitian population had little or no access to electricity—a predicament that undermines the nation's agricultural sector, which is largely unable to process and store its produce.[49] In a unique pilot project driven by a community engagement process, EarthSpark International, a nonprofit that markets small-scale solar products, determined that farmers were prepared to prepay for limited electricity in the same way that they purchased cell phone minutes.[50]

In Les Anglais, a small agricultural town in southwestern Haiti, sunset meant cessation of work and reliance on kerosene lighting or candlepower. Working with local officials and nongovernmental organizations, the Haitian government, Electricité d'Haiti (the national electric utility), and others, EarthSpark found a way to take advantage of a bit of local providence: Digicel, the regional provider of cell phone service, had erected a large telecommunications tower, and the small diesel generator that powered the tower yielded more electricity than the tower required. In the spirit of collaboration and experimentation, Digicel agreed to let EarthSpark tap into the excess capacity to launch a small pilot microgrid capable of serving 54 customers. In addition to providing street illumination at night, the microgrid enabled each household, through its "Spark-Meter" (figure 2.8), to enjoy 3 W LED lighting and a duplex outlet for charging small

Figure 2.8
EarthSpark International's SparkMeter (smart meter) installed in Les Anglais, Haiti. Photo Credit: Daniel Schnitzer, courtesy of EarthSpark International.

appliances and cell phones.[51] Thus, the cascade of waste energy from one service supported another service, demonstrating the resource-sharing principle that is central to infrastructural ecology.

Thanks to a $1.1 million grant from USAID, EarthSpark has been able to expand to a fully solar-powered smart grid that as of 2015 served 440 homes and businesses in Les Anglais. Proceeds from paying customers cover operational costs and will help finance development of additional microgrid initiatives.

Reciprocity across Service Sectors

In ecology, *mutualism* refers to mutually beneficial interactions between two distinct species. In the field of agroforestry, for example, trees and shrubs are planted alongside herbaceous plants, in spatial arrangements that yield economically and ecologically beneficial interactions.[52] In the examples highlighted in this section, the metaphor of reciprocity is applied to urban services—such as water storage and energy production—that interact across sectors.

Capitalizing on Energy and Water Reciprocities in India

With about 22 percent of its citizens lacking access to the electric grid and an equal number hobbled by unreliable power,[53] India has sustained a strong appetite for coal and gasoline. Today, approximately 19 million diesel engines pump India's irrigation water from shallow wells, emitting millions of tons of CO_2.[54] Meanwhile, given India's latitude and an annual count of 300 clear, sunny days, the nation's theoretical solar power reception is about 5,000 trillion kWh/yr, a small fraction of which, if captured effectively, could meet the entire country's power requirements.[55]

In January 2010, the government of India's Ministry of New and Renewable Energy launched the Jawaharlal Nehru National Solar Mission, with its short-term goal being to increase the nation's solar capacity to 10 GW by 2017. This goal has been revised twice, and as of June 2015, it was set at 100 GW. Of this, 40 GW was expected to come from rooftop solar and 60 GW from large-scale grid-connected solar projects,[56] making India one of the global leaders in solar energy.[57] To achieve this goal, the ministry planned to reduce the cost of solar power generation, primarily through enabling policy and research and development. It is working with private sector investors to finance a rapid rollout of photovoltaic power plants.[58] The solar program is being financed, in part, by doubling the levy on fossil fuels (coal, peat, and lignite) while reducing tariffs on the imported materials needed to develop solar and wind installations.[59]

India is successfully reducing barriers to large solar installations, in part by eliminating the real estate costs associated with new solar farms. To avoid the machinations of opportunistic landholders—some of whom have increased prices by 20 percent to 30 percent on vacant property earmarked for solar use by the state—the government has begun to position solar farms above bodies of water,[60] a strategy that also protects land that is critical for agriculture. Moreover, both capital costs and the per-unit costs of energy production are about 15 percent lower than for land-based projects.[61]

In a global first, India's 17,700 km (11,000 mi) network of irrigation canals is being topped with solar arrays. One such array, inaugurated in April 2012, is a 1 MW photovoltaic farm that sits atop a 750 m (2460 ft) section of the Narmada Branch Canal near Chandrasan village in Gujarat State. The array was built by the Gujarat State Electricity Corporation with the support of Sardar Sarovar Narmada Nigam Ltd (SSNNL), which developed, owns, and operates the canal network (figure 2.9).

The solar project pays rent to the owner of the body of water, and, as a side benefit for SSNNL, the shade from the panels yields an estimated annual savings of 900 m^3

Figure 2.9
One megawatt solar PV power project atop the Sardar Sarovar Namada Nigam Limited (SSNNL) Canal near the city of Kadi in Gujarat State, India. Photo Credit: Hitesh vip, Wikimedia Commons.

(237,750 gal) of water that would otherwise have evaporated in the arid heat.[62] The Gujarat State Electricity Corporation enjoys a reciprocal benefit in that the proximity of the cool thermal mass of water near the solar panels helps reduce their temperature, improving their efficiency and extending their life span. According to one study, these floating arrays could achieve performance improvements of 10 percent or more.[63]

As of early 2014, SSNNL was investing in a second stage: a 10 MW solar power plant along a 10 km (6.21 mi) stretch of the same canal. The expenditure incurred for a 1 MW pilot project was over Rs 170 million ($2.55 million; 1 crore = 10 million Indian rupees), but the cost of panels has since come down, and the estimated cost for the new project is Rs 110 million to 120 million ($1.64 million to $1.79 million) per megawatt.[64]

The state-run National Hydro Power Corporation (NHPC) is constructing a 50 MW floating solar power station on a reservoir in Kerala in southwestern India. The station is modeled on Japan's 60 MW Kagoshima Nanatsujima Mega Solar Power Plant, which was constructed as part of a broader effort by the Japanese government to move away from nuclear power. The Japanese facility consists of floating platforms—with a total area of 1.27 million m^2 (13.7 million ft^2)—atop the Nishihira and Higashihira irrigation ponds in Kato City, Hyogo Prefecture.[65] (In addition to reducing evaporation and increasing efficiency, the panels also suppress the rate of algae growth in still water.) India's floating platforms will be anchored to ensure stability during bad weather. The project's cost is estimated to be between $64 million and $72 million, and construction on the initial pilot plant began in October 2014. Because of the increased efficiency of arrays that have been installed over water, another entity, the Kerala Airport Authority, is considering installing its own power plant over a nearby body of water.[66]

The coupling of energy and water is not without its own difficulties. There are concerns, for example, about whether electricity can always be safely transmitted from the floating platforms, and about the potential corrosion of the panels (as a consequence of exposure to moisture).[67] But, to date, the benefits have been shown to outweigh these risks.

Multipurpose Dams: Water Storage Potential in Viet Nam, Liberia, Korea, and Jordan

Many emerging economies are located in regions that are subject to extreme oscillations in the hydrologic cycle—too much rain followed by too little. And many such economies are also hobbled by energy poverty. One option for addressing both water management and energy production is the multipurpose dam, which combines water storage with energy generation and can also incorporate a range of additional functions, from flood control to biodiversity enhancement, improved inland navigation, and recreation. Through flood regulation and increased water storage, in particular, such dams can also help respond to climate instability.

Despite their advantages, multipurpose dams (as well as hydro dams in general) are not without controversy. There is ongoing debate, for example, about the environmental damage associated with the construction of access roads and transmission networks. And, although it is possible to reduce emissions by clearing vegetation before an area is flooded, the decomposition of submerged vegetation may increase greenhouse gas emissions (methane in particular). Operationally, there may also be conflicts between different needs. For example, peak energy demand may require drawdowns at times when farmers need to store water for future use.[68] There is also the risk that technical problems in one part of the project could impact another (e.g., if the generator fails, it may be impossible to pump water). Finally, disparate but interdependent functions may lead to more complex (and therefore more costly) operations and maintenance. Multipurpose dams do hold promise, however. Two examples—one in Viet Nam and the other in Liberia—show the potential for collateral benefits and improved rates of return compared with single-purpose reservoirs.

The Trung Son hydropower project, a medium-sized facility under construction in a remote mountainous village in Thanh Hoa Province in northwestern Viet Nam, will provide 260 MW of renewable power while avoiding approximately one million tons of CO_2 emissions annually. A 25 km (15.53 mi) access road to the village and its surrounding areas is another dividend.[69] Built with financial support from the World Bank (at an estimated cost of $410.68 million), the dam—which has a storage capacity of 112 million m^3 (146.49 million yd^3)—will provide flood control and supply irrigation for agriculture.[70] Water impoundment is scheduled to begin in October 2016.

In Liberia, only about 10 percent of urban residents and less than 2 percent of rural dwellers have access to electricity, apart from that provided by privately owned generators. At the same time, Liberia's renewable water resources per capita are among the highest in Africa. A small-scale "run-of-the-river" project currently under way on the Mein River in the Suakoko District will reduce energy poverty through a local distribution grid serving about 2,000 households, as well as government buildings and industry; it will also supply water for domestic, livestock, and irrigation purposes. As an added bonus, local agencies will be trained to operate and repair the mini hydropower facility and the minigrid. Thus, in addition to fostering learning by doing and building local capacity, the project is expected to help boost the market for minihydro as a form of distributed-energy system.[71]

Because of its small scale, the project largely avoids greenhouse gas emissions and may circumvent other problems—such as environmental damage—that have been associated with medium- and large-scale hydro. Moreover, the technology involves no displacement of local inhabitants. Located in a rural area of the country, northeast of the capital, Monrovia, the 1 MW hydropower station was cofinanced by USAID and the United Nations Industrial Development Organization at a cost of $1.8 million.

The idea of recovering energy from existing reservoirs is not new. In most cases, single-purpose dams can be converted into multipurpose dams by retrofitting turbines onto existing storage structures. A recent example is Korea's transformation of its freshwater procurement strategy—a barrage across a tidal estuary, Lake Sihwa—into an energy-producing project. By 2011, mounting industrial pollution had rendered this reservoir unusable as a source of fresh water. The barrage was therefore refitted with tidal turbines, and it is now one of the largest tidal power plants in the world. In addition to producing 552 GWh of power annually, enough to serve 200,000 homes,[72] the project features a wetland park, a visitors' center, and highway and rail links between two previously remote locales.

Just as single-purpose dams can be converted into multipurpose ones, other water services can be transformed into small hydropower plants, optimizing the use of the resource. In nearly any water infrastructure facility—a drinking water supply plant or wastewater treatment plant, for example—underutilized pressure from water discharge can be integrated into small-scale hydropower systems, offering opportunities for energy recovery.

In 2008, at a wastewater plant in Samra, Jordan, the heads (dammed up bodies of water) from both intake and discharge into the Duleil Wadi were outfitted for energy production.[73] Similar energy-recovery mechanisms can be integrated into navigation locks by using the difference in water levels during the filling and emptying of the locks. Finally, at desalination plants, where water is passed at high pressure through semipermeable membranes to separate out dissolved salts (the reverse osmosis process), residual liquids with high salt content are still at high pressure. These can be discharged through a turbine to recover a portion of the compression energy used at the outset of treatment.[74] Findings from a series of case studies suggest that wherever head pressure can be found in water infrastructure, there is an opportunity to generate electricity.[75] This suggests that as a matter of best practice, water infrastructure should be designed to incorporate hydro turbines.

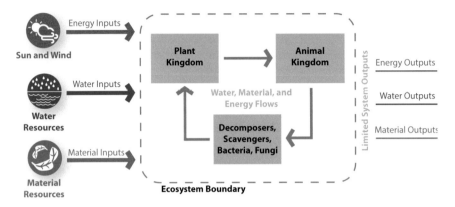

Figure 2.10A
Diagram of flows of energy, water, and matter in an ecosystem. Credit: Hillary Brown and Byron Stigge.

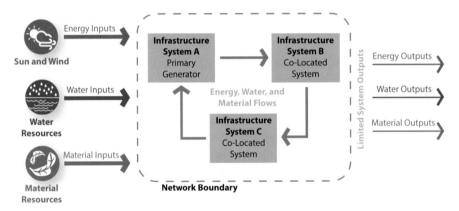

Figure 2.10B
Concept diagram of an infrastructural ecology. Credit: Hillary Brown and Byron Stigge.

Integrating Multiple Systems: Toward a Circular Economy

Herbert Girardet, a consultant, professor, and author who works in the field of sustainable development, has observed that "for cities to become sustainable, they must develop circular metabolisms."[76] When a constructed system (figure 2.10B), whether an industrial park or an entire city, has a circular metabolism, it mimics the patterns that can be found in complex living systems (figure 2.10A); in particular, closed-loop material flows, energy cascades, and the regeneration of natural assets. Two points deserve special emphasis: in such circular economies (the term that will be used here), (1) waste is valued *as a resource*; and (2) the recovery of energy and material resources yields extra services, thereby creating added value—that is, more value per unit of resources expended.

During the many decades of postwar economic exuberance, most residents of urban areas, at least in industrialized nations, had little awareness of how resources were derived or what became of the wastes they produced. Thus, the legacy infrastructure of this era is largely single purpose and linear rather than multifunctional and closed loop. Some emerging economies, however, have maintained links to traditional holistic practices and habits of mind and may thereby be better positioned to unlock the value of the circular economy.

The examples in this section demonstrate how the synergistic coupling of different infrastructural services can support the urban metabolism while reducing resource intensity. In this way, with adroit planning and coordination across sectors, infrastructure can support economic growth while avoiding resource depletion and ecosystem degradation. The convergent thinking that underlies such initiatives is the backbone of infrastructural ecology and is the first step toward a circular economy.

Chinese Polyculture: Flood Control and Food Production

Polyculture—the combination of agriculture and aquaculture—is an ancient practice that maximizes the economic use of both land and water. It is embedded in the cultures of South Asia and Southeast Asia and has been practiced in China since the seventeenth century.

During the settlement of the marshy Pearl River (Zhujiang) Delta four hundred to five hundred years ago, increased population limited suitable land for farming, and other resources were scant.[77] The area was also subject to recurrent flooding. As both a water control measure and a means of food production, farmers constructed dikes across small tributaries, forming ponds. Additional ponds were dug to drain the marshes, and the excavated soil was used to build more dikes.[78]

Typically, the ponds were relatively narrow rectangles that conformed to the topography. The strips of land between them were used to cultivate rice and were fertilized by nutrient-rich silt from the ponds. The farmers also cultivated carp—a hardy and adaptable fish that was readily introduced into the constructed ponds. (Initially food for the elite, carp became a staple; in fact, the Chinese word for "fish" means "surplus.")[79] Because carp can withstand crowding in contained areas, as many as four species were purposely cultivated together. Arranging themselves according to their preferred habitat, they made use of the same ecological system: the grass carp fed on vegetation near the shore;

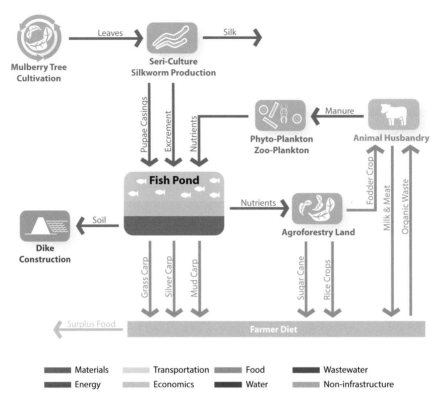

Figure 2.11
Chinese polyculture (dike-pond system), Zhujiang Delta, China. Credit: Hillary Brown and Byron Stigge.

the bighead carp and silver carp ate the zooplankton and phytoplankton that circulated at midwater; and the mud carp fed off bottom-dwelling plants and animals.[80]

On the excavated soil adjacent to the ponds, the farmers established silkworm farms, where silk fiber was extracted from the cocoons of the worms. The worms fed off cultivated mulberry trees, and their excrement and pupae (the casings discarded when the worms metamorphosed into insects) were introduced to the ponds as supplemental fish food. Carp feces, in turn, provided natural fertilizer for the rice crops. Livestock was also a link in this circular economy. The manure of the pigs, ducks, and chickens that were raised on the dikes fertilized the fishponds, adding nutrients that encouraged the growth of plankton, on which the fish fed.[81] Figure 2.11 illustrates the reciprocities between flood protection, food production, and the silk economy.

Grounded in generations of empirical knowledge, this form of polyculture was extremely productive, maximizing the use of land and water while recycling organic waste as nutrients. By conserving and recycling resources, the farmers not only made good use of their space and time but also enjoyed a diverse diet with marketable surplus. Today, in some parts of the Zhujiang Delta, the circular economy of the dike-pond system still prevails.

In Chinese polyculture, trees, silkworms, fish, fowl, pigs, and fodder produced a diverse diet and drove an economy, all without consequential waste leaving the system. This symbiosis contrasts with modern production, which typically operates according to a linear (take-make-consume-waste) path. It also illustrates the preindustrial disposition to perceive the world in terms of connectedness—a holistic mindset that survives today in the term *systems thinking*. Whereas the industrialized world emphasizes the optimization of constituent parts, systems thinking values the whole and perceives that it is larger than the sum of its component parts. Thus, the focus in systems thinking is on the interactions between those parts and on the relationship between any given system and other, larger systems.

Infrastructural Symbiosis at Work: Tianjin, China

Founded in 1984, the Tianjin Economic-Technological Development Area (TEDA)—a large, multifaceted industrial area south of Beijing—was one of 14 economic development zones assembled under the aegis of China's State Council, the macroeconomic management agency of the National Development and Reform Committee (NDRC). In 2001, under the auspices of the State Environmental Protection Administration (SEPA), which had established a demonstration program for ecoindustrial parks the previous year, TEDA launched a pilot ecoindustrial park within its development zone, located adjacent to the Tianjin port on Bohai Bay. The SEPA program had been earlier inspired by methodologies outlined in *Industry and Environment,* a 1997 publication (in Chinese) of the United Nations Environment Programme, which described environmental management of industrial estates.[82]

In 2005, the NDRC selected TEDA to demonstrate a model circular economy. Firms from four major industries, including many multinational companies, subsequently colocated at TEDA's development zone. The industries—automotive, electronics, biotech and pharmaceutical, and food and beverages—are all served by a shared public infrastructure system, as are the commercial and residential areas included within TEDA's 45 km^2 (17.37 mi^2) boundary. (Such an arrangement is not uncommon in Chinese ecoindustrial parks;[83] in fact, Chinese ecoindustrial parks also typically include areas for scientific research, business, and services, as well as for industrial production.)[84]

The Bohai Bay site, previously a saltpan, had poor, saline soil, and the closest reservoir was 52 km (32.31 mi) away. The scarcity of land and water, along with the need to provide low-cost utilities, spurred the use of unconventional infrastructural

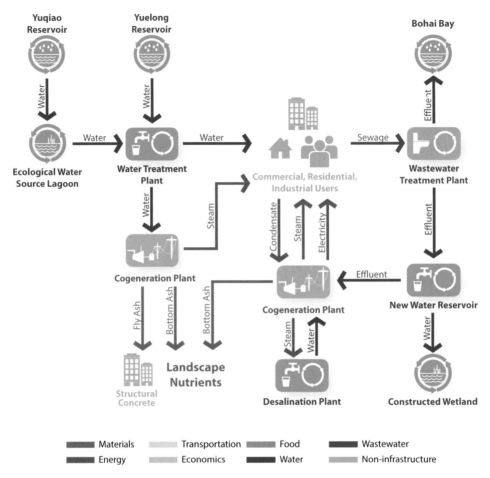

Figure 2.12
Diagram of Tianjin Economic-Technological Development Area, China. Credit: Hillary Brown and
Byron Stigge.

technologies. To maximize efficient use of water and natural resources, TEDA gradually
evolved a symbiotic arrangement of utilities and industry, which included water and
energy cascades; wastewater treatment and recovery and sludge reuse; desalination;
and management of solid waste (see figure 2.12).

   The scarcity of local groundwater and the distance to the reservoir drove investment in a water-recycling plant capable of supplying the park's diverse manufacturing processes. Water reclaimed from the TEDA Sewage Treatment Plant is subjected to
reverse osmosis and microfiltration. Sixty percent of that water is dedicated to use as

process water, and the rest is used for landscape irrigation or to recharge an artificial wetland.[85]

Energy cascades are a major feature of the power sector: several coal-fired cogeneration plants provide industrial users with electricity, hot water, and process steam. Some of the process steam is used as an energy source in a nearby desalination facility. Condensate from large users of steam—such as auto manufacturers—is returned for reuse as cooling water for the nearby power plants run by Tianjin Binhai Energy Development Company. TEDA's 120 GWh waste-to-energy plant is another power source. Bottom ash, a residue of combustion from the plant's furnaces, is used for floor-tile manufacturing.[86]

A beneficial by-product of the utility's water and energy cascades is the manufacture of new soil, which is helping to remediate the excess salinity of the topsoil in the farmland around TEDA, thereby conferring environmental and economic benefits. The soil is being created from a combination of Bohai sea sediments, caustic soda sludge (waste from the industrial processes and wastewater treatment plant), and fly ash from the power plants.[87] Solid waste exchanges among the companies include the recycling of scrap metals (especially aluminum and lead), glass, oil, and silver (from the electronics cluster). Scraps from food processing are used by nearby farms, and nutrients from biotech and pharmaceutical processing are converted into organic fertilizer.[88]

As of 2013, there were 87 symbiotic exchanges in TEDA's industrial network. Overall, the infrastructural synergies accounted for 89,355 metric tons (98,497 tons) of CO_2 abatement, 321,076 metric tons (253,925 tons) of landfill diversion, and an overall revenue increase of CNY 110,400,000 ($17,664,000).[89] The networking among the utilities is relatively local: the average distance between energy-related exchanges is 2.7 km (1.67 mi), and the corresponding figure for water-related exchanges is 3.5 km (2.17 mi).[90]

TEDA's careful itemization of exchanges, aggregate savings, and benefits provides valuable support for the advancement of infrastructural symbiosis; this is in contrast to the still incomplete inventory maintained by Hammarby Sjöstad, which was described in chapter 1. Other distinctions between the two examples are also instructive. The genesis of each enterprise was quite different. Hammarby Sjöstad was established as a mixed-use village but was not designed around the needs of industry, whereas TEDA's primary intention was to attract and promote industry while simultaneously providing utilities that would serve the area's industrial, commercial, and residential developments. TEDA's ecoindustrial initiative was regulated by national, local, and facility-level standards and guidelines. In contrast, Hammarby Sjöstad took shape through negotiated consensus and was implemented through public-private partnerships and market-driven approaches.

Although both TEDA's top-down and Hammarby's bottom-up approaches succeeded in implementing closed-loop solutions, each model has followed a different trajectory. As there appear to be no replications of such strategies at a comparable scale to date, Hammarby Sjöstad may prove to have been a one-off accomplishment in the industrialized West. This is not the case in China, where industrial symbiosis and its infrastructural corollary have long been embedded in official development goals.

China appreciates the importance of clusters as spaces for experimentation and innovation, and as a strategic means of moving development along a more sustainable path.[91] As early as 2002, the central government had accepted the concept of the circular economy as a viable strategy for promoting rapid economic growth while avoiding energy and resource shortages.[92] In 2009, China released a policy decree, Notice on Promoting the Development of a Low Carbon Economy in National Demonstration Eco-Industrial Parks. As of 2013, 85 ecoindustrial parks had been established.[93]

Ecoindustrial parks are only one of the mechanisms China is deploying to promote circular economies. SEPA, which provided guidelines for the development of circular economies, has also promoted ecocities, which can be viewed as circular economies at a macro scale. Whereas ecoindustrial parks focus on industrial production, ecocities are designed with both production and consumption in mind.[94] Three factors—rapid urbanization, the shift toward a market economy, and a desire to avert the environmental consequences of congestion and the social consequences of unequal resource distribution—have enabled the ecocity concept to take hold in China.[95] From the meso level of the ecoindustrial park to the macro level of the city, China is using effective recycling of resources to reshape its industrial, infrastructural, and urban regimes, and to decouple resource consumption and carbon emission from economic growth.

Forward Thinking

Infrastructural ecology offers a planning and design framework that promotes beneficial exchanges across service sectors—reducing system costs, improving performance, and lowering social and environmental impacts. Patterned on the workings of natural ecosystems, infrastructural ecology fosters synergies through colocation or other forms of networking. As will be illustrated in the following chapters, networked facilities are ultimately more sustainable because they integrate the free work of natural systems; reduce carbon intensity and greenhouse gas emissions; mitigate and adapt to a destabilizing climate; and enhance the communities in which they are situated.

3

The Soft Path: Aligning Water Infrastructure with Natural Systems

Let not even a drop of rain water go to the sea without benefiting the world.[1]
—Parakrama Bahu the Great, AD 1153–1186

Emerging economies face vast and enduring infrastructural deficiencies. Too much high-quality water is being extracted and used for nonpotable purposes. Many regions are subject to extreme oscillations in rainfall, and unless excess water can be stored, livelihoods are at risk. As populations increase (and urbanize), demand for water is expanding rapidly, intensifying long-standing inequities affecting women and children, who have historically been burdened with the responsibility for transporting water. Continued deforestation and the proliferation of surfaces impervious to water are disrupting runoff patterns, causing harmful floods. In the absence of adequate regulations, industrial and human waste is causing mounting environmental damage. Finally, climate instability is intensifying water scarcity, affecting the closely bound interactions between water, energy, and food production and exacerbating competition for scarce resources.[2]

In most parts of the developing world, the need for safe water and sanitation far outstrips the available ecological—not to mention sociopolitical and financial—capacity to construct conventionally designed facilities. And where such facilities already exist, too many regions are beyond their reach. In the coming years, single-purpose, "hard-path" solutions—such as dams; aqueducts; groundwater mining; and large, complex, centralized means of moving, storing, and treating water—will not suffice to meet burgeoning demand. To serve the global billions who lack sustainable access to safe drinking water, basic sanitation, and sufficient water to support livelihoods, next-generation water infrastructure will require more holistic approaches to water management. These include "soft-path" systems, which (1) work according to nature's logic; (2) integrate natural processes into supply, conveyance, treatment, and reuse; and (3) can be more readily adopted for decentralized applications.[3]

The case studies that follow offer examples of multiple-use systems, whereby single water sources serve health, sanitation, and agricultural needs. They also illustrate time-honored practices for securing and storing water in dry regions; controlling surplus and scarcity through green infrastructure; and using soft-path approaches to purify waste, recover nutrients, and obtain fuel for energy production.

Multiple-Use Water Systems

The two models described in this section feature sophisticated hydraulic engineering, multiple-use systems, and soft-path treatments. Each illustrates how a single water source can be configured to satisfy domestic, hygienic, and livelihood needs. These constructions, which exemplify the logic of natural systems, may prompt a rethinking of the complex, centralized, industrial-era approaches to water infrastructure that have, among other consequences, encouraged profligate water consumption. Many emerging economies still have the opportunity to create localized water systems that are more holistic and less complex.

"Benefiting the World": Cascading Systems in Ancient Sri Lanka

In the undulating, seasonally desiccated landscapes of Sri Lanka's dry zone, the ancient Sinhalese civilization created one of the most sophisticated irrigation systems of the ancient world. The waterworks occupied great swaths of land, in the northern and eastern sections of the island, that lie in the central highlands' rain shadow.[4] This unique water storage and distribution network stretched across multiple watersheds, encompassing about 40,000 km² (15,400 mi²),[5] and supported both upland farming villages and the palaces of the kings who governed the region's river basins (figure 3.1).

The network—a system of water tanks, manmade (dammed) lakes, and ponds, which branched off from perennial rivers and streams—irrigated over one-third of the available agricultural land downstream.[6] Formed with artificial earthen embankments (known as *bunds*), the tanks were positioned at oblique angles to the water's trajectory to reduce the impact of floating debris. Each tank also had its own small catchment zone for monsoonal rainwater collection.

By transporting both river water and rainwater downstream to the next tank in a linear sequence, the catchment system allowed water to be efficiently reused by the next downhill settlement. The system also provided unusual resilience in the face of droughts and flash floods, which would otherwise have degraded the topsoil.

Figure 3.1
Upper Walawe basin, Sri Lanka. Credit: K. Shannon, 2007, redrawn from G. Foster, "The Concept of Regional Development in the Indigenous Irrigation Systems of Ceylon," *Yearbook of Association of Pacific Coast Geographers* 31 (1969): 91–100.

Tanks situated along the tributaries were interconnected, forming a cascading arrangement in which a tank was typically fed by both runoff and groundwater infiltration from the tank immediately above it in a continuous pattern (see figure 3.1). The smallest tank, at the top, was spring fed. Some of the tanks in the middle, forested areas, which supported slash-and-burn agriculture, were shadowed by large trees, reducing evaporation and keeping the water cool. Other medium-sized tanks in this zone stored monsoon water for villagers to use during the dry season to irrigate rice paddies as well as nonrice and perennial crops in communal uplands. The largest tanks, at the bottom, served the agricultural plains below.[7]

Each village tank was collectively owned and managed. Tank water served domestic and livestock uses and supported an irrigable area of about 42 ha (104 acres). Directly uphill from the tank, constructed wetlands, consisting of aquatic plants, filtered the drainage water from the paddy fields, reducing any pollutants.[8] Water outlets admitted the cooler bottom water directly to the village, which was typically situated below the tank, at the base of its dam wall. The hydrostatic pressure from the tank served to raise the water table in the lower agricultural areas and fisheries beyond. The water then flowed into the "buffalo wallow" area (figure 3.2). Overall, this cascading system allowed stored water to be used, fed back into the system, and used again several times over.

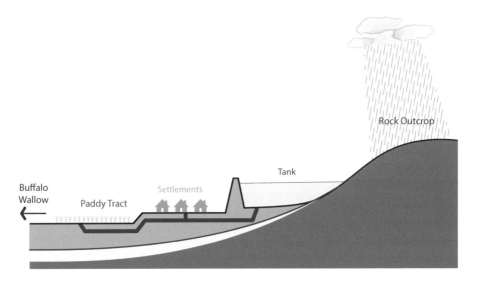

Figure 3.2
Diagram of a paddy tract, conceptual section. Credit: Logman Arja, redrawn from W. Gooneratne and S. Hirashima, *Irrigation and Water Management in Asia* (New York: Sterling Publishers, 1990), 154.

Sri Lanka's unique hydraulic arrangements were governed by a set of well-defined management practices—including the payment of water taxes—at both the local and kingdom levels. The system provided a foundation for livelihoods from the fifth century BC to the thirteenth century AD. By the end of the fifteenth century, much of it had fallen into ruin.[9] Nevertheless, the model remains exemplary: an integrated, recycling water commons that addressed both upstream and downstream needs and was wrought in accordance with natural systems, attentive to limits, and supported by collaborative planning and operation.

A Multipronged Utility: Water Regulation in Cistercian Abbeys

During the twelfth and thirteenth centuries, Europe's Cistercian monks located their abbeys in forest clearings adjacent to rivers, building on terraces or constructing levees to protect the compounds from flooding. The monastic communities had a gift for collective water management and practiced a form of "multiple water use" that is enviable today.

Because their livelihood as agriculturists and animal breeders depended on a reliable water source, the monks—using calculated gradients—dug ditches and laid conduits to divert river water into the abbey grounds. Routed through underground pipes made from hollowed-out tree trunks, ceramic, or lead, the water passed first through sedimentation basins and then to a central distribution point and a separate dispensing fountain, where it served not only the monks but also their guests and lay workers (figure 3.3).

Greywater—the water that had been used for drinking, heating, cooking, bathing, and laundry, as well as in the breweries and for liturgical purposes—was emptied into the main wastewater channel, where, supplemented by rainwater, it irrigated the gardens before passing into fishponds, known as "stews." (Many abbeys bred fish to supplement their diet.) As the greywater passed further downstream, it was joined by blackwater from the latrines and the hospital. The combined wastewaters then filled a series of reservoirs, where power from the water's release drove various workshops—including mills for corn and wheat for the brewery—and supported other industrial needs.[10] After domestic, agricultural, and industrial needs had been satisfied, the wastewater rejoined the riverbed downstream.[11] The sequential flow of the multiple uses helped maintain relatively good hygiene.

Like the Sri Lankan tank system, the Cistercian monks' approach to water use and reuse is highly relevant today. As water sources become increasingly unreliable or inaccessible, we must learn to use the same water stream numerous times before discharging it to receiving waters and out to sea. And this new paradigm—single stream, multiple uses—will depend on more widespread application of soft-path systems.

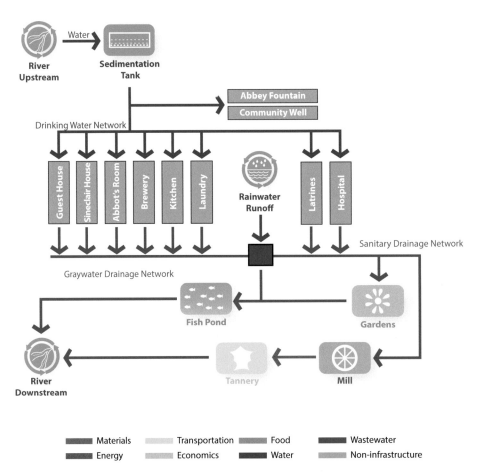

Figure 3.3
Freshwater supply and wastewater discharge in Cistercian monasteries. Credit: Byron Stigge, redrawn from C. J. Bond, "Mittelalterliche Wasser-versorgung in England und Wales," in *Geschichte der Wasserversorgung*, Bd. 4, ed. Frontinus-Gesellschaft e.V. (Mainz: Verlag Philipp von Zabern, 1991), 149–182.

Capture and Storage for Water Sufficiency

Throughout the world, in urban and rural areas alike, rainwater harvesting is currently being promoted as a means of counteracting unsustainable water practices. In Japan and South Korea, for example, stored rainwater is used in firefighting and as an emergency source during power disruptions. In many developing countries subject to seasonal water shortages, indigenous water-harvesting methods were abandoned when centralized Western models were established. Today, however, in parts of the Global South, some traditional techniques are being revived.[12]

Brazil, China, and India have created enabling legislation and subsidies to promote harvesting.[13] In the semiarid northeastern region of Brazil—which is subject to dry months and where climate change forecasts call for worsening future conditions—the Program for 1 Million Cisterns has constructed more than 230,000 cisterns since 2008 through a combination of private and government funding.[14] In 18 of India's 28 states, rooftop and cistern systems are now mandatory for new buildings.[15] As a result, cities such as Mumbai and Bangalore have been better able to endure water shortages.[16] Pakistan installed its first urban rainwater system in 2010. From the huge roof of the Faisal Mosque complex in Islamabad, rainwater—as much as 3 million l (792,500 gal) per day during the rainy season—is piped to a nearby forest aquifer, where it is filtered through sand to recharge the city's water table. What would otherwise have been runoff ends up as clean drinking water.[17] As water-conscious nations develop proficiency in collecting, storing, and treating rainwater, such technical improvements can promote approaches to localized water sourcing that are even more cost-effective.

Lost Practices Regained in Microcosm: Harvesting Rainwater in Sri Lanka

By the late twentieth century, as Sri Lanka was undergoing urbanization, industrialization, and deforestation, the indigenous system of cascading tanks described earlier in the chapter had largely been abandoned thanks to nationalization of these water resources. This change ended their communal ownership and maintenance, and, given limited government resources, the system deteriorated.[18] Meanwhile, Sri Lanka's groundwater supply increasingly was being tapped for irrigation and other commercial uses, and contamination from pesticides and fertilizers had increased dramatically. As of 2009, one-third of the population was without a piped water supply and relied instead on manual pumping from wells. In 2010, the International Water Management Institute predicted that, by 2025, annual per capita water availability would fall to 1,700 m^3 (about 450,000 gal)—the threshold of water scarcity.[19]

Although rainwater had been harvested from roofs for many centuries, the practice had become obsolete as the population increasingly relied on springs and wells. Especially in the dry zone, once this supply was exhausted, many residents—typically women or children—had to make several trips each day to access water, often traveling more than two miles on foot or bicycle.[20]

At the time of writing, domestic rainwater harvesting—a microcosm of Sri Lanka's sophisticated tank system—has been successfully reintroduced. Each domestic system consists of a 7.5 m^3 (nearly 2,000 gal) tank—enough to provide a family of five with

Figure 3.4
Rainwater harvesting tank in Sri Lanka's Northern Vavuniya District. Photo Credit: Passana Gunasekera, USAID. Courtesy USAID, https://www.usaid.gov/results-data/success-stories/ancient -water-scheme-give-sri-lankans-new-lease-life.

at least 20 l (5.28 gal) of water per day throughout the dry period (institutional tanks hold 10 m^3, or nearly 2,600 gal). The freestanding, lobular tanks are made of either ferro-cement (figure 3.4) or brick, materials that can easily be accessed locally. The system relies on a simple mesh with charcoal and gravel filters, which removes the first flush (initial surface runoff) and eliminates dust and debris.[21] The harvested rainwater is generally safe to drink (the tanks are sealed to prevent bacteria and insect breeding). In urban areas, rainwater is used to supplement piped potable water, a practice that, according to one study, can reduce water bills by more than one-third.[22] Moreover, the tanks help to control urban flooding. Any excess that spills from the tanks is directed to pervious areas to recharge aquifers.

Harvesting began in the mid-1990s, when the government, with assistance from the World Bank and USAID, created the Community Water Supply and Sanitation

Project (CWSSP) to assist with upgrades to water supply facilities. Further investments in research and technology fell under the auspices of a nongovernmental organization (NGO), the Sri Lanka Rainwater Harvesting Forum.

By 1998, the National Water Supply and Drainage Board had constructed 73 systems in one district alone; more than 31,000 systems can now be found throughout the country. Although most are in rural areas, where they have been implemented through government projects or by local NGOs, many are in urban locations. In 2005, the government passed a national policy rendering rainwater harvesting mandatory for certain new building types in urban areas.[23]

In 2012, a $483,000 USAID grant helped institutionalize rainwater harvesting in rural Sri Lanka through demonstration projects, workshops, and training.[24] Under the grant, the Lanka Rain Water Harvesting Forum (LRWHF) provided training for masons to erect 750 household tanks. Community participation in construction helped offset high initial costs, which might otherwise have prevented many households from adopting the technology. The system has improved overall sanitation and reduced diarrheal and other waterborne diseases. Time saved by household members can be dedicated to other activities, such as increased agricultural cultivation to strengthen food security and generate income.

Renewing India's Hydrologic Lifeline

In India, which depends on the monsoon season (typically June to September) for nearly three-quarters of its annual precipitation, water is revered as a purifying and restorative element in ritual and prayer.[25] Although blessed with many rivers, the nation is characterized by large spatial variations in the distribution of water, which intensifies the temporal variability of the resource.[26] Historically, engineered water infrastructure— including rainwater capture and storage—has been a lifeline and has been closely tied to village and religious life. This connection gave rise to the incomparable architecture of India's centuries-old *stepwells* and *baoli*, highly ornamented subterranean edifices constructed between the second and eighteenth centuries, which provide year-round access to groundwater-filled pools. These constructions doubled as public monuments and places of congregation, and they were often found adjacent to temples (figure 3.5). However, during the colonial era, which introduced widespread pumped irrigation, the stepwells and *baoli* were allowed to deteriorate. Although some recent initiatives to refurbish some of these landmarks have been made, few have succeeded—echoing the fate of the more modest engineered structures on which Indian agriculture once depended.

In the drought-prone Godavari River basin, which spans almost the entirety of India's midsection, rivers flow seasonally, and groundwater is unreliable.[27] To ensure

Figure 3.5
Chand Baori, in the village of Abhaneri near Bandikui, Rajasthan. Photo Credit: Doron, Wikimedia Commons.

the availability of water for the entire year, villages traditionally relied on a system
of large tanks (*cheruvu*) and small tanks (*kunta*) to impound water during the torrential
storms of the rainy season.[28] The tanks—artificial reservoirs positioned across a slope—
were often organized in a cascading pattern not unlike those in Sri Lanka. They were
sited to take advantage of local depressions, and the villagers then built up linear mounds
to complete the enclosures.[29] Typically, check dams—small constructions across the
waterway—were used to reduce flow velocity and encourage the settlement of sediment.

In the semiarid Maner basin (a subbasin of the Godavari), all 24 microbasins relied
on these artificial reservoirs to irrigate water-intensive crops, including maize, rice, and
cotton. Villages also relied on tank water for bathing, watering cattle, and washing
clothes (figure 3.6). Tank management practices were fundamental to the local culture.
For example, members of lower castes were charged with removing built-up silt from
the tanks,[30] and women left layers of marigold flowers in the tanks to "clean" them.[31]

Averaging 10 ha (about 25 acres) in size, these constructed natural systems harbored
their own microecologies, providing vital ecosystem services, which included nutrient recycling, groundwater recharge, and flood mitigation. The tanks also helped to

Figure 3.6
Tank Lalbagh, Bangalore, India. Photo Credit: Cyphor, Wikimedia Commons.

support immediate habitat—a diversity of trees and plant life—and allowed the culti-vation of fish.[32] When the tanks were low or empty, they sprouted grass, which could then be harvested for roof thatch or cattle fodder. Silt mined from the tanks was used as fertilizer for the fields.[33]

Beginning in the late 1940s, while India was still under colonial rule, and later under India's state government departments, local decision makers were displaced from tank management. Maintenance effectively ceased, and weeds grew unchecked. In the nearly 60 years since independence, the government has made haphazard efforts to restore the tanks but has failed to take account of basic tank ecology and hydrology.[34] Inflows decreased, and silt, which has been eroding from the now deforested river basin, rapidly accumulated, dramatically reducing the tanks' storage capacity.[35] Meanwhile, the state government supplanted traditional hydrological regimes with groundwater extraction, borewell drilling, and the construction of hydro dams, canals, open irrigation ditches, and other structural means of water diversion. More recently, formerly self-sufficient villages have seen their livelihoods disappear, farmers have migrated to major urban areas, and suicide rates among farmers have increased.[36]

Climate instability has intensified water shortages in the Godavari watershed. Farms are receiving less total rainfall, and summer monsoons are bringing more intense and more frequent downpours.[37] The area's population is expected to nearly double by 2050, with water needs following suit.[38] According to several estimates, by 2050, India will need another 120 major storage dams to meet projected needs, but such construction would displace populations, damage ecosystems, and come at great financial cost.[39]

From 2004 to 2006, the World Wildlife Fund supported a pilot project to revive tanks in the Maner subbasin. Undertaken in concert with the NGO Modern Architects for Rural India (MARI) and relying on community labor, the project was designed to showcase the economic viability of micro-restoration—as opposed to state-sponsored macroinfrastructural projects—and to identify the social and policy tools that would be needed to scale up such an effort.[40] At a cost of $103,000—nearly three-quarters of it in the form of in-kind labor from local farmers—66,224 metric tons (73,000 tons) of sediment were removed from 12 tanks that served an 11 ha (27 acre) area with a population of 42,000 people.[41] The performance of the tanks improved: irrigation capacity greatly increased, and reliance on groundwater withdrawal ceased.[42] Significantly, the restoration of the water tanks can also increase groundwater recharge during high-rainfall years, which can then be drawn on in times of prolonged drought.

The indirect benefits and by-products of the restoration—both agricultural and ecological—helped recover project costs quickly and even provided a profit. Cessation of groundwater pumping decreased electrical costs. Because tank water is fed by gravity, the increased water tank volumes created additional pressure, allowing the water to flow farther and to reach more farms.[43]

The clay silt that the farmers removed was rich in nutrients and carbon, and was ultimately used by nearly 900 farmers to fertilize 600 ha (1,482 acres). The nutrient excavation also reduced the need for inorganic fertilizers and improved soil quality, strengthening the plants against crop pests and lowering pesticide use while improving overall yield.[44] According to the farmers' findings, the augmented topsoil increased soil moisture retention from four to seven days, benefiting the crops.[45] To reduce the cost of silt removal, the farmers had created 16 silt and soil islands in the wetlands, resulting in an environmental benefit: the recovery of avian and fish stock density and diversity in the water tanks. Finally, this low-energy, soft-path solution increased carbon fixation throughout the area, a beneficial climate mitigation measure.[46]

The economic returns—valued overall at Rs 5,850,500 ($91,267)—came primarily from the increased yields on the 50 ha (123 acres) amended with silt, where the production of groundnuts and maize provided the biggest gains.[47] (This figure excludes the increases in milk production made possible by the greater availability of fodder, and the increases in fish production made possible by the larger volumes of water in the tanks.)[48]

Historically, tanks had been maintained by the community; weeding, the removal of silt for soil amendment, and other tasks were integrated into sociocultural routines. In the World Wildlife Fund pilot, village tank committees, which had ceased to function, were reconvened, and wages were offered to some for silt excavation. For the landless poor, earnings from the farms' higher yields and the increase in fish stocks compounded the wage benefits. The restored functionality of the tanks also supported the reemployment of a former class of washer (laundry) men. Finally, increased local employment reduced migration to urban centers.[49]

The restitution and enhancement of natural and constructed systems yielded a complex, integrated system of benefits: improved climate resiliency, rejuvenated agriculture, refurbished social structures, and revivified local employment. Restoring the small-scale, zero-energy tanks was revealed as a practical way to implement *Gram Swaraj*—village self-determination—which Mahatma Gandhi had envisioned as instrumental to a democratic India.[50]

Scaling up tank restoration could help reduce India's need for large-scale water projects. Approximately 208,000 village water tanks remain across the country, representing a low-cost, low-tech, locally meaningful approach to achieving a sustainable water supply. Projections have shown that if the existing tanks in the Maner subbasin were to be desilted by approximately 4.572 m (15 ft), they could store as much as 2.94 billion m^3 (780 billion gal), meeting water demands even during low-rainfall years.[51] Such a decentralized and low-impact initiative stands in contrast to the massive water infrastructure currently under consideration, such as the $4 billion Polavaram Dam under construction in the lower Godavari, which is likely to displace more than a quarter of a million people and inundate 60,000 forested hectares (148,826 acres).[52]

In late 2014, the state of Telangana (formerly part of Andhra Pradesh) undertook Mission Kakatiya (named in remembrance of the rulers who initially developed the system), which will entail the systematic revival of 46,531 tanks over a five year period, at a cost of Rs 20,000 crore ($320 million). Among the objectives of the project is the restoration of community-based irrigation management as a means of increasing the incomes of small farmers; to this end, social organizations for tank management will be revived. The project will also include the cultivation of high-value, less water-intensive crops; fisheries development; and improvements to the health and diversity of livestock.[53] To fund the work, villages, nonresident Indians, nonprofits, and corporations are being encouraged to "adopt" tanks for restoration; donors' names will be inscribed on plaques.[54] The program can be viewed as both an instrument of decentralization and an example of participatory management reforms, both of which are critical to the community contributions that are the focus of chapter 6.[55]

Green Infrastructure at Work in Emerging Economies

Both pluvial (rainwater) and fluvial (river water) flooding pose major threats to many urbanizing parts of the globe, including those that experience dry seasons. Even where annual precipitation is low, torrential tropical storms can instantly inundate both flat and low-lying terrain (whether it is steep or flat). The low-income communities that occupy unsuitable but affordable sites in poorly drained floodplains are particularly at risk. In the wake of floods, survivors may suffer lingering aftereffects, including disability and disease from exposure to contaminated standing water.[56]

The term *green infrastructure* refers to natural or seminatural (human-modified) systems that rely on soil and plants to reduce erosion and flood intensity while amplifying ecosystem services.[57] The implementation and maintenance of green infrastructure depends on regional and community-level collaboration—both formal and informal. Typical strategies include reforestation; afforestation (the establishment of new forest cover); the conservation of wetlands, coastal dunes, marshes, and reefs; and the construction of new wetlands and riparian buffers. Many recently adopted urban strategies, such as green roofs and permeable pavement, are also classified as green infrastructure. In addition to the provision and strengthening of ecosystem services, such nonstructural management solutions can improve health, biodiversity, and agricultural yields. Perhaps most important, they can lead to new employment and training, which are among the keys to building local capacity in fragile states.

Despite being an essential and effective expression of socioecological resilience, such innovative responses to threatened urban environments are rarely considered—let alone valorized—in cost-benefit studies. Moreover, because the substitution of "green" for "grey" water infrastructure has been a mainstream practice for only a few decades, perceived risks are higher, and projects under consideration must meet more exacting standards.

A Flood-Control Alternative: Urban Parkland in Curitiba, Brazil

Curitiba, an industrial metropolis situated in the upper Iguaçu River basin of Brazil, was founded on a forested floodplain crisscrossed by multiple rivers. During the city's dynamic growth phase—between 1940 and the early 1990s—poor land-use policies and weak enforcement along the banks of the Iguaçu River, as well as those of its upper tributaries, brought recurrent flooding to the dense and burgeoning settlements on the floodplain. Unrestricted urbanization enlarged the floodplain's impermeable areas. Streams were diverted into underground channels, which, along with obstructions like bridges and landfills, made peak discharge impossible to manage. Flooding increased rapidly throughout the 1950s and 1960s.[58]

Protecting against 50- or 100-year floods by conventional means would have required deepening the river along a 50 km (31 mi) corridor or constructing costly dams or levees.[59] But, in 1972, while other Brazilian municipalities were using federal flood-mitigation funding to channelize rivers and construct dams, Curitiba took a far more cost-effective route. Under special federal legislation, the city used the funding to transform part of the Iguaçu floodplain into an urban park system, preserving the valley floors and the profile of the river basin as a flood-control area. Planted areas—including protection strips to help control erosion—were built as buffers along streams and were further stabilized by extensive tree planting.[60]

Despite these precautions, two major floods—in 1983 and 1995—inflicted nearly $100 million in combined damage.[61] In 1996, with funding from the World Bank, the city's environmental sanitation program undertook a series of radical control measures, which included designating extensive parkland (in the main zone of the floodplain) to serve as river storage and establishing a metropolitan-scale master plan for urban drainage. Curitiba's comprehensive, nonstructural approach to flood management fashioned new protection areas and linked existing ones (figure 3.7). The park system, which tracks the main river through the urbanized area, ranges from 300 m to 1 km (0.2 to 0.6 mi) in width and covers a 20 km^2 (7.72 mi^2) area.

To reduce the potential increase in peak flow as the areas along upstream tributaries underwent development, the urban drainage master plan incorporated linear parks along the tributaries. In addition, a metropolitan-area investment fund was established to enable upstream counties to implement their own drainage plans.[62]

Legislation, including the Municipal Decree for Riverside Areas Preservation, established subcatchment (tributary) easements and riverbank preservation to curtail flooding.[63] Curitiba's Zoning and Land Use Law of 2000 ensured the continued preservation of the valley bottoms—designating four environmental preservation areas where development was restricted or prohibited, and allowing owners to transfer development rights.[64]

Figure 3.7
View of Curitiba, Brazil, from Parque Barigui, Curitiba, Brazil. Photo Credit: Marco Guerra, Wikimedia Commons.

A parallel channel to one side of the river was dug, which serves both to demarcate the limits of allowable settlement and to accommodate increased river flow as needed.

The greenways lining the river transformed Curitiba: green space ratios increased from less than one m^2 (10.76 ft^2) per person to 64 sm^2 (689 ft^2),[65] and parklands now make up 20 percent of the urbanized area.[66] (The average for American cities is 8.1 percent.)[67] Adding to this green infrastructure are more than 1.5 million trees—"green lungs" that volunteers have planted since the 1970s along the city's streets and avenues. The profusion of green infrastructure—which features 35 parks, over 1,000 conservation areas, 78,000 m^2 (about 840,000 ft^2) of natural forest, over 145 km (90 mi) of trails, numerous deep lakes, and flood-resilient playgrounds—has improved citizens' health. Because it serves as a buffer zone, the parks network also preserves and enhances Curitiba's extraordinary biodiversity: the city hosts nearly 4,000 species—an exceptionally high number for a large, circumscribed metropolitan area.[68]

The Parques Lineares de Fundo de Vale (Linear Parks of the Bottom of the Valley), funded almost exclusively by federal flood-control funds, have become primary recreational areas that include restaurants and other concessions that finance park maintenance. The property values adjacent to the parks, which appreciated significantly, contribute increased tax revenues to the city.[69] According to the United Nations Environment Programme, the cost of transforming flood zones into planted parks and artificial lakes—including the cost of relocating the favela residents—was approximately one-fifth the cost of building concretized channels.

Curitiba's unique flood-management plan was the first of several cross-sector innovations executed by the Institute for Urban Planning Research (IUPR), a visionary team of architects and planners established in 1965, whose 14-member board cut across most city administrative divisions. One of the IUPR's founding members, architect Jaime Lerner, eventually became three-time mayor of Curitiba (and later governor of Parana). Beginning in the 1980s, Lerner's administration, to make way for the enlarged riverside park, began to remove some of the burgeoning favelas (informal settlements), rehousing the population in high-rise buildings constructed along new public-transit corridors. These novel bus rapid-transit corridors link the city along radial axes, incorporating tree planting along dedicated bus lines, pay-as-you-enter subway stations, and parallel side streets for local traffic. Lerner's overarching mission of social inclusion is perhaps best personified by waste-for-transit privileges, through which the poor can earn ridership in exchange for collecting trash. Few infrastructural models since have surpassed the combined accomplishments of Curitiba's IUPR.

Solving for Scarcity and Surplus: Indonesia's Versatile *Biopori*

Despite Indonesia's more than 5,590 rivers and an annual rainfall that varies from 1,650 mm (65 in) in Jakarta to 2,260 mm (89 in) in Medan in north Sumatra,[70] as of 2011, just under half the population enjoys improved sewage disposal,[71] and 55 percent enjoys access to safe, piped water.[72] Each year, widespread water pollution leads to waterborne diseases—including diarrheal illnesses, cholera, and typhoid fever—among nearly one-third of the population, and to 50,000 deaths. The socioeconomic burden of illness and loss of life heavily depresses the country's gross domestic product (GDP).[73]

In urban areas, local government is responsible for water services. Bogor City, in the West Java province, is one of the fastest-growing municipalities in the country. At least 80 percent of the city's residents lack access to safe water, relying instead on purchased water or on river water polluted by sewage, garbage, and industrial waste.[74] Excessive extraction of groundwater has also damaged soil biodiversity, destabilized soil structure,

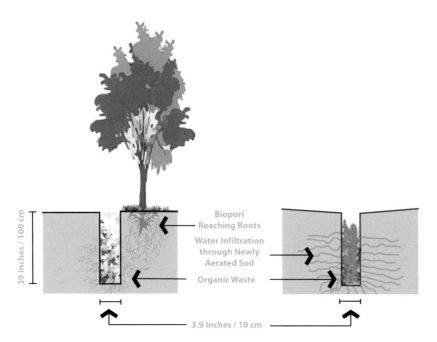

Figure 3.8
Sectional diagram of biopore. Credit: Logman Arja, redrawn from http://www.basicknowledge101 .com/photos/2015/biopore%20holes.jpg.

and contributed to widespread environmental damage. Although Bogor City enjoys an average annual rainfall of 4,500 mm (177 in)—and is in fact nicknamed "Rain City"— extensive building coverage and compacted soil cause routine flooding, and the city's rivers, both big and small, are much degraded by pollution.

In 2004, Kamir R. Brata, a researcher at the Bogo Institute of Agriculture, developed a low-tech approach to improving water drainage that simultaneously resolves other problems. Vertical holes between 10 and 30 cm (3.9 to 11.8 in) in diameter and 100 cm (39.4 in) deep are hand-bored into the ground at intervals of 100 cm (39.4 in). The empty holes are then filled with organic material—such as leaves, grass clippings, and organic waste from kitchens and gardens (figure 3.8). These "biopores" (*biopori* in Indonesian) replicate the natural process whereby water rapidly infiltrates undisturbed soils to great depths. Typically, a concrete, metal, or plastic collar is used to stabilize the biopores where they meet the ground.

The holes and their organic fill attract earthworms, and as the worms access and digest the waste, their burrowing regenerates the soil, and their tunnels increase soil porosity.

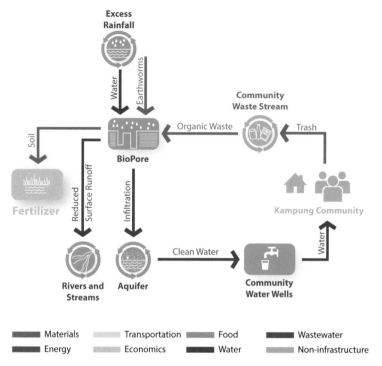

Figure 3.9
Operational diagram of biopore. Credit: Logman Arja and Hillary Brown.

Before the holes are refilled with new organic material, rich compost is extracted for use as fertilizer.

By encouraging biological activity, the *biopori* renew soil fertility. At the same time, they allow infiltration of excess stormwater at a rate of about 3 l (0.79 gal) per minute. Infiltration through soil and rock helps purify the water before it is extracted from local wells. Along with other, larger-scale groundwater-recharge wells—up to 5 m (16.4 ft) deep—the decentralized, smaller *biopori* help reduce the risk of land subsidence by replenishing the water table. Finally, newly acquired composting habits have helped reduce the amount of strewn garbage, adding to the other cobenefits (see figure 3.9).[75]

The program's cooperative approach is key to its success. Since 2007, an estimated 1,000 biopores have been put in place. Although most have been financed through local government initiatives (one biopore costs about $20), planning, implementation, and maintenance are participatory.[76] Thanks to workshops and training campaigns conducted in schools and private institutions, the technology has spread widely, increasing resilience to annual flooding.[77]

Community participation through in-kind labor has empowered local populations to take ownership of biopore programs. In 2014, as part of World Environment Day, 15,000 residents of Probolinggo, a city in East Java, dug 15,000 *biopori* in more than 600 neighborhoods.[78] The World Bank's Water and Sanitation Program, which sponsored the initiative, required that village heads obtain broad community agreement to participate. Thanks in large part to the nationwide biopore campaign, Indonesia plans to achieve universal access to clean water by 2019.[79]

Synergistic Solutions: Tree Cover in the Humbo District, Ethiopia
Urbanizing areas find that new water regimes impact existing flora and fauna—and often entire ecosystems.[80] Between 1995 and 2000, Ethiopia lost much of its original high forest cover; today, less than 2.7 percent remains (figure 3.10A).[81] In Humbo, a mountainous administrative district in the Great Rift Valley in southwestern Ethiopia, an unusual, farmer-managed reforestation initiative improved groundwater quantity and quality, stemmed flooding, improved water quality, increased recharge, and made it possible to meet needs for potable water, all while realizing maximal environmental and socioeconomic benefits.

Within this district, increasing population density led to the swift conversion of what had once been mountainous woodlands to agricultural use, including intensive grazing. As forests were cleared for timber, firewood, or charcoal production, the topsoil on the hillsides eroded and became subject to periodic landslides, creating downstream gullies and covering valuable farmland with silt, cobblestones, and boulders. The majority of Ethiopia's 80 million people rely on agriculture for their livelihoods; it accounts for 50 percent of GDP and 80 percent of employment. The land devastation created by deforestation has led to food insecurity, poverty, and famine.[82]

Since its inception in 2006, the Humbo Assisted Natural Regeneration Project has restored 2,728 ha (6,741 acres) of lost forest—an undertaking that has not only rejuvenated damaged ecosystem services but is also expected to sequester 970,360 tons (880,296 metric tons) of carbon equivalent over its 60 year lifespan.[83] Sponsored by World Vision Australia and World Vision Ethiopia (the implementing office of the international NGO), and funded through the purchase of carbon credits by the World Bank's BioCarbon Fund, the Humbo initiative is the first large-scale forestry project in Africa to be registered under the UN's Clean Development Mechanism. Under that program, the project is permitted to sell certified emissions reductions (CERs); over the project implementation period, revenue from CER sales will produce an income stream of more than \$700,000;[84] and most of the total project cost of \$1,379,303 will be financed by CERs. An anticipated 338,000 tons of CO_2 equivalents are to be sold by 2017, of which 165,000 tons have already been purchased by the World Bank's BioCarbon Fund.[85] As part of the project, the area will be harvested after 12, 22, and 31 years, with each harvest leaving 50 percent of the standing biomass to support ecosystem services. No carbon will be expended in logging, as animals will be used to transport all forest products. Carbon sequestration and other benefits will be carefully monitored.

The reforestation project entailed the development of a new model of community land use that is suitable for wide replication in other Ethiopian regions (figure 3.10B).[86] In 2006, when the project was initiated, the first step was to protect areas that still had living stumps or remnant trees by closing them to users. Unlike conventional reforestation approaches, which use costly nursery stock, the farmer-managed natural regeneration strategy applied in Humbo relies largely on encouraging new growth from still-living stumps (the roots of trees cleared as long as 60 years ago can develop shoots). To support the farmer-led restoration of this particularly poor area, World Vision Ethiopia held workshops and training sessions. Whereas farmers had previously slashed this regrowth in order to plant crops, they were now being trained to nurse the stems, pruning them for faster growth.[87] Supplemental planting of both indigenous and naturalized species from nurseries began the following year.

Seven separate community cooperatives, located adjacent to the reforestation site, were involved in the project. Each cooperative developed a plan and bylaws for the use and management of the lands, which were designated as community holdings, and for the first time, titling and land-use rights were legally recognized. The communities earn shared income from some of the CER sales, as well as from the sale of timber products and by harvesting grass from designated woodlots. The carbon revenues are also being invested in activities that will support food security, including the construction and maintenance of local infrastructure.[88] Most significant, the self-managed restoration—and the international recognition that came along with it—restored local pride.[89]

In addition to its social and economic benefits, the project created significant water catchment, reducing the sediment runoff that had previously been affecting the water quality of Lake Abaya, which is downstream of the restoration area. The project has also replenished the springs and subterranean streams that support the region's water supply. Among the other environmental benefits are ecosystem services from the forest, such as increased vegetation cover and improved microclimate. Finally, by restoring the connections between fragmented forest resources, the program revived a refuge for local and migratory species, which had been lost to deforestation. Building on the success of the forest cooperatives, whose existence is now embedded in Ethiopian law, the

Figure 3.10A
Aerial view of Ethiopia's Humbo District Reforestation Initiative (before). Photo Credit: © 2015 Google Digital Globe.

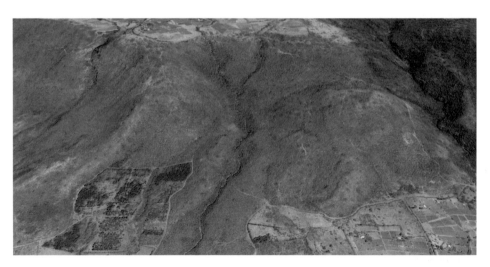

Figure 3.10B
Aerial view of Ethiopia's Humbo District Reforestation Initiative (after). Photo Credit: © 2015 Google Digital Globe.

Ethiopian government has set a goal of restoring 15 million ha (37.6 million acres) of degraded and deforested land by 2025.[90]

Water Reuse and Nutrient Recovery: Sustainable Imperatives for the Anthropocene

Although centralized wastewater treatment systems have been installed in developing nations, a significant number have been associated with multiple failures, possibly because of insufficient funding and a lack of engineering expertise.[91] Most rapidly urbanizing locales, however, have yet to centralize sewers and treatment systems, creating an opportunity to implement decentralized technologies from the start. Such systems can be low energy, and, significantly, both constructed and natural systems can be beneficially integrated into the urban fabric. The choice of technology should be based on context, ideally relying on readily available resources.[92] Most important, the systems should not only cleanse water for reuse but should also provide residual returns, whether in the form of nutrients, potential energy sources, or both. Two exemplary cases follow.

The Enduring Synergies of the Kolkata Wetlands

In many parts of the globe, marginalized populations still rely on urban wastewater for irrigation, despite the risks to human health and the environment associated with this practice. In India in particular, conventional sewage treatment plants have had abysmal maintenance and performance records.[93] In the East Kolkata Wetlands (EKW), however, a low-cost, zero-energy treatment option decreases direct pollution of the receiving waters, reduces the need for chemical fertilizers, and provides livelihoods for nearby communities. When properly planned and implemented, linking wastewater management with food and energy production offers many opportunities to safely close both the water and nutrient loops.

The EKW is perhaps the oldest and largest integrated water treatment and resource recovery practice in the world. Known as "the kidneys of Calcutta," the EKW is situated on the southeastern fringe of that enormous city. A natural asset that supports diverse uses and benefits, the 12,500 ha (30,888 acre) area is a unique mosaic of bodies of water, vegetable fields, canals, fishponds, roads, and villages. Forty-six percent is wetland, 39 percent is agricultural, and the rest is made up of settlements and municipal garbage-disposal sites.[94] Used for aquaculture since the 1850s, the area's brackish wetlands have served continuously as the city's major sewage treatment plant since the 1930s while also providing food products for Kolkata and livelihoods for a periurban community that would otherwise be impoverished. Aspects of the EKW approach, which was originally devised by

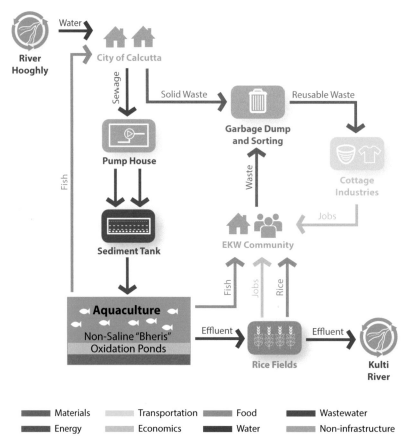

Figure 3.11
Diagram of East Kolkata Wetlands, Kolkata, India. Credit: Byron Stigge and Hillary Brown.

farmers and fishermen, suggest an alternative route for the development of sanitary engineering in emerging economies (figure 3.11).

The sewage treatment ecosystem, which has both natural and constructed elements, consists of salt meadows, settling ponds, oxidation basins, and intertidal marshes. Water quality management relies on natural indicators.[95] Each day, roughly 600 million l (150 million gal) of Kolkata's wastewater flows, via underground sewers, to an area where it is pumped through a sedimentation tank into open channels that terminate in *bheris*, 254 nonsaline fishponds (figure 3.12) that are natural settling ponds that act as solar reactors.

The sewage is detained in the ponds for days, where it undergoes biodegradation (biochemical dissolution of organic material by microbes). Aquatic flora and fauna trap

Figure 3.12
"Bheris" seen beyond women on a footpath in the East Kolkata Wetlands, Kolkata, India. Photo Credit: © Brett Cole.

the freed chemical energy, using it to support their own growth. The degraded organic material is then taken up by local plants (water hyacinth and duckweed), which extract heavy metals and also shade the fish and plankton in the summer months. The organic material also provides food for fish and, along with the plant life, oxygenates the habitat. The healthy level of dissolved oxygen, in turn, supports the biodegrading microbes.[96]

This comprehensive nutrient recovery cycle (the recapture of nitrogen, phosphorous, potassium, etc.) provides an annual supply of about 11,000 metric tons (12,125 tons) of edible freshwater fish, mostly carp and tilapia—approximately one-third of the city's fish consumption.[97] About 12 different species of fish are raised in the various ecological niches of the pond ecosystem.[98]

As the partially cleaned effluent flows by gravity from the sewage-treating fishponds, it irrigates the surrounding paddy fields, producing about 15,000 metric tons (16,534 tons) of unmilled rice each year. Studies have shown that because of the high nutrient content, the rice irrigated by the effluent is both of higher quality and more profitable than that from lands cultivated with conventional irrigation.[99]

The city's garbage-disposal site is also located at the water treatment wetlands, where about 2,500 metric tons (2,756 tons) of municipal solid waste arrive annually by truck. After women and children (known as ragpickers) have separated out the recyclables—such as scrap metal and plastic—for cash, the remaining waste is left to decompose and becomes natural compost for agricultural use.[100] As part of the same complex, in the areas near the garbage dump and around the *bheris*, local households rent small plots for sustenance and income, growing vegetables directly on organic waste. Each day, approximately 150 metric tons (165 tons) of produce are transported to the city.[101]

The environmental benefits credited to the EKW, valued at $38 million per year,[102] include water purification, oxygen production, biodiversity, and productivity (with respect to both the food chain and biogeochemical cycles), and the wetlands also serve as water-storage basins during the monsoon season. Finally, in addition to supporting the cultivation of fish, rice, and vegetables, the EKW sustains other cottage industries, including the production of pulses (edible seeds), oils, paper pulp, and thatching materials.[103]

In 2003, under the Ramsar Convention, India designated the EKW as a "wetland of international importance." Since then, conservation of the site has been under the auspices of the East Kolkata Wetlands Conservation and Management Authority. However, 93 percent of the functional areas of the EKW are privately owned; the remaining 7 percent are under the control of state government or farmers' cooperatives. Maintenance, much of which is handled by farmers and fishermen, entails dredging the ponds of silt, harvesting the water hyacinths, and making and repairing fishnets. Relying on traditional knowledge and practices that have come down over generations,[104] farmers and fishermen undertake pond preparation, fertilization, fish stocking, and fish harvesting. Today, the EKW supports approximately 20,000 families.

Until the late 1990s, the ecosystem and its associated economy enjoyed relative stability. Kolkata's rapid growth, however, has sparked conflict between those who support the conservation of the EKW and those who favor redeveloping the land for residential, commercial, and industrial uses. Additional siltation, stemming from advancing urbanization, has also taken a toll by reducing flows. Meanwhile, increasing levels of industrial pollutants, including heavy metals and medical waste, are entering the urban sewage system, posing new threats to human health and species biodiversity. Rising sea levels, which increase the infusion of saltwater into the wetlands, also pose a risk. Nonetheless, the EKW demonstrates the viability of an arrangement in which natural systems successfully perform the otherwise energy- and chemical-intensive services of conventional wastewater treatment. Equally significant is the integration of traditional knowledge into the design and management of this unique, informal system.[105]

Such an approach to wastewater reuse offers both opportunities and challenges for urban populations in the developing world. To date, there have been few replications at this scale of the EKW's low-cost, culturally appropriate, decentralized technology in highly urbanized contexts. One obstacle to replication is the need to prevent industrial wastewater, which is incompatible with food production, from mixing with urban sewage. Properly planned and monitored, however, comparable systems could be established in similar climatological zones.

Energy from Sewage: Southern Africa and Chile

According to the U.S. Environmental Protection Agency, as of 2010, methane emissions from municipal wastewater systems accounted for 6 percent of global methane emissions.[106] The use of anaerobic biodigestion at wastewater treatment plants, a carbon-neutral technology that can recover wastewater methane at an urban scale, has become common in Europe (Germany has 6,800 large-scale anaerobic digesters)[107] and, to an increasing extent, in the United States. Some European municipalities (and now New York City) have also begun to divert food waste from landfills to supplement the supply of waste sent to the digesters. Since anaerobic treatment takes place in closed vessels, odors are contained; it is therefore practical to colocate waste treatment with other land uses.

Aside from a few examples, such as the EKW, the significance of both water reuse and resource recovery has been undervalued in the sanitary sector, and replication has been limited. Sewage contains both thermal and chemically bound energy, both of which can be used beneficially. In cooler climates, for example, thermal energy—which is derived from indoor building temperatures—can be extracted close to the source and can be used (by means of heat exchangers) to preheat water. More significant for developing nations in tropical and subtropical climates, however, is the chemical energy of wastewater, which, through anaerobic digestion, can be transformed into biogas, a fuel consisting mostly of methane. Biogas can be used for cooking or heating, as well as to generate electricity, and it can also be upgraded to vehicle fuel.[108] Thus, it is now feasible, in certain climates, for wastewater processing systems to become net energy producers—a prospect rendered even more appealing by the fact that the systems in question can be maintained by less skilled operators than for conventional systems.[109] (Yet another advantage is that biodigestion—the process that converts wastewater to biogas—considerably reduces the overall volume of biosolids, decreasing the cost of sludge disposal.)

The practice of biogas recovery is becoming much more widespread in emerging economies and is being undertaken at various scales—from households to institutions, neighborhoods, and even municipalities. Perhaps the most widely disseminated technology is the small, fixed-dome biogas anaerobic digester, which, as of 2004, served an estimated 15 million households in China (which has begun to scale up and industrialize production of the digesters); that same year, more than 3.5 million units were installed in India.[110]

In 2010, in the African nation of Lesotho—where roughly half the inhabitants lack sanitary facilities and a piped water supply—more than 140 household-sized, fixed-dome digesters were installed as on-site sanitary solutions, closed-loop systems that process wastewater at the point of generation while providing a local source of energy. The design, which is based on concrete or brick domes, is adapted to locally available materials and skills and satisfies both technical and cultural requirements associated with the handling of human waste.[111]

Built by Technologies for Economic Development, an organization based in Maseru, the capital of Lesotho, the below-grade digesters treat blackwater and greywater, as well as kitchen waste (and, in some cases, animal waste).[112] The waste enters the dome through an external inlet.[113] Then, within the digester, anaerobic bacteria decompose the organic material, reducing the overall volume of the biosolids.[114] Biogas, a by-product of this process, is extracted through a separate gas outlet. Readily replacing costly bottled gas, the biogas provides 100 percent of cooking energy in summer and about 20 percent to 40 percent in winter.[115] Slurry, another residual of the natural processing, can be used for both irrigation and fertilization of vegetable gardens—increasing domestic food production, enhancing the moisture-retaining capacity of soil, and inhibiting the germination of weed seeds.[116]

The La Farfana Wastewater Treatment Plant, which treats about 60 percent of the wastewater in Santiago, Chile, is one of Latin America's largest treatment plants. The facility opened in 2003 with eight digesters, each with a capacity of 15,000 m³ (3.96 million gal). Only 12.5 percent of the biogas produced, however, was used to heat the digesters; the rest was flared. In 2009, an upgrade performed by the French company Degrémont (which was partially financed through the sale of carbon credits under the UN's Clean Development Mechanism) made it possible to clean and condition the remaining 87.5 percent of the biogas (to remove hydrogen sulfide, trace oxygen, and nitrogen), which is now piped to the Metrogas Town Gas Plant, located 13.7 km (8.5 mi) to the west of Santiago. The biogas cleaning and conditioning systems were a new technology for Chile. Today, the treated gas supplies some 35,000 homes; more significantly, the project—which cost $3.5 million—has reduced annual CO_2 equivalents by 26,340 metric tons (29,034 tons).[117]

Worldwide, almost one billion people live without access to toilets; according to a UN report, if that sanitary waste were collected for biogas production, the electricity generated—valued at $200 million per year—could serve 10 million households.[118] Still, the use of biogas at the scale of wastewater treatment facilities faces both real and perceived barriers. In developing countries, key obstacles are lack of capital and lack of adequate operations and maintenance expertise. The Clean Development Mechanism, however, offers one means of investing in renewable energy technology that is capable of harnessing the latent energy in municipal wastewater.

Heading Down the Soft Path

Soft-path solutions, which are attuned to all aspects of the hydrologic cycle, provide a key means of achieving water balance—that is, approximate equilibrium between water input and output—in urbanizing areas. As the examples in this chapter illustrate, such systems can be less costly and have a lighter footprint than conventional, hard-path technologies.[119] The water harvesting now practiced in Sri Lanka and the village tank systems once endemic to India are effective, low-tech, decentralized technologies that can complement centralized solutions—and in some cases replace them. According to the United Nations, which views such technologies as critical to combating water conflicts in many regions of the world, the biggest obstacle to rainwater harvesting is the lack of policies and legislation.[120]

Urban drainage and flood-protection solutions incorporating naturalized, low-impact collection areas—like those used in Brazil, Indonesia, and southern Africa—can improve health and strengthen communities while securing future water resources. The monumental task of improving sanitation for those who lack it can be undertaken through distributed systems that recover nutrients for agricultural reuse. Promising approaches, including anaerobic digestion, not only facilitate wastewater reclamation and reuse but also yield biogas for energy production.

Gains are being made in developing new, and resurrecting old, solutions to water treatment in urbanizing locales. Ultimately, however, progress in this arena must go beyond technological solutions. To ensure proper implementation and ongoing operation, institutional, legal, and regulatory arrangements that are consonant with social and cultural conditions must be developed.

Post-Carbon Infrastructure:
Power, Heat, and Transport

A useful analogy is to see traditional societies as relying on instantaneous (or minimally delayed) and constantly replenished solar income, while modern civilization is withdrawing accumulated solar capital at rates that will exhaust it in a tiny fraction of the time that was needed to create it.
—Vaclav Smil, "World History and Energy," in *Encyclopedia of Energy*

At various points during previous millennia, innovations marking the transition from animate energy to the "inanimate prime movers"—water and wind—diffused throughout China and Persia, on to Rome, and then to the rest of Western Europe.[1] These new technologies capitalized on energy flows induced by gravity or gravitational tides, and on the breezes unleashed by the effects of solar radiation, which were then captured by windmills. They were based on an understanding of two phenomena: first, that the velocity of a liquid in motion can be transferred to wheels fixed on a horizontal axis; and second, that the horizontal pressures of air movement can be harnessed to produce rotational power.

Renewable hydro and aerodynamic energy sources continue to stimulate our imaginations today. The many innovative examples in this chapter demonstrate that local renewable energy resources, creatively linked with other utility sectors, can help decarbonize power and heat production along with transport while realizing other benefits.

Emerging Economies and the Carbon Challenge

In developing countries, the push to reduce carbon emissions is often viewed as secondary to meeting needs for basic services, education, and economic development. To the extent that resources directed toward the development of low-carbon infrastructure

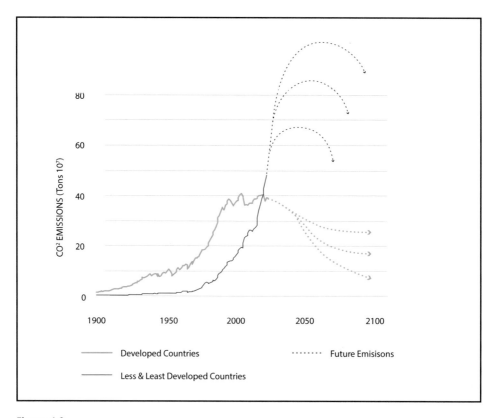

Figure 4.1
Global greenhouse gas emissions for developed countries and developing countries, 1950–2050.
Credit: Carbon Dioxide Information Analysis Center, Oak Ridge National Laboratory.

are perceived as compromising short-term economic development, efforts to implement low-carbon solutions will be an uphill fight. But where low-carbon infrastructure
can be demonstrated to not only support multiple infrastructure needs but also foster
long-term economic development, there is hope of coupling economic growth with
low carbon emissions.

The United Nations Sustainable Development Goals—the successors to the Millennium
Development Goals—refer only tangentially to carbon emissions and set no hard targets
for reduction.[2] One reason is that carbon emissions in developing nations are rising dramatically and are projected to continue to rise for the next few decades.[3] But total annual
carbon emissions in developed countries seem to have peaked and are on track to begin
falling in the coming decades, as figure 4.1 shows. Thus, the focus on low-carbon strategies for urban expansion and economic development should be in developing countries.

For the 77 countries designated by the United Nations as "least developed," average annual greenhouse gas emissions per capita are 3.0 tons[4] per person—just under 25 percent of the average for G20 countries, which is 11.6 tons per person.[5] These figures reflect a vast gap in carbon intensity that can be seen in nearly every sector of energy consumption—from overall built space to lighting, heating, cooling, and use of appliances. Meanwhile, many developing nations have a rapidly growing middle class, which naturally aspires to the same levels of health and comfort as their counterparts in industrialized nations.

According to one estimate, if the 5.8 billion people in 145 developing economies rapidly achieved the carbon emissions intensity of G20 countries—through, for example, car-based transportation, meat-based diets, and carbon-intensive buildings—annual global carbon emissions would rise from 34 gigatons to 82 gigatons—a 240 percent increase.[6] Such an upsurge would have catastrophic effects on global temperatures and climate patterns. The challenge, then, is to increase the quality of life in emerging economies without impairing ecosystem services or interfering with essential activities, such as food production.

In developing nations, hundreds of thousands of small-scale projects are needed to supply energy for cooking, lighting, and small appliances. This is the first step toward economic growth, as well as improved health, education, and shelter. Particularly for the 2.8 billion rural poor in developing countries, energy poverty precludes reading, working, or learning at night; in fact, enabling children to read and study through the evening hours may be the single greatest global benefit of power generation in poor rural areas. Most important, energy must be supplied at costs that are manageable, given existing economic constraints.

To overcome this daunting challenge, energy services must be reexamined from a broader, systems perspective—one that touches on every aspect of energy supply, delivery, and efficiency, as well as human behavior. In practical terms, this amounts to a switch in focus. Instead of approaching energy needs and sources in isolation, a systems approach considers the *services* that energy provides and then asks how best to provide those services—within the context of low-carbon growth.

Essentially, low-carbon development entails reducing both total and per person (or per dollar of GDP) greenhouse gas emissions while supporting economic growth. Although, for example, turning off the lights in an office building may save energy, it is not a "low-carbon approach" if it reduces workers' ability to do their jobs.[7] Carbon reductions made at the expense of service quality are not actually carbon reductions but rather are service reductions. And it is one of the premises of this book that the level and quality of infrastructure services in emerging economies can and should be *increased.* Ideally, low-carbon investments are developed through a fully integrated plan-ning process that relies on culturally appropriate means of accommodating growing urban populations while fostering improvements in productivity and quality of life.

This chapter highlights opportunities in low-carbon, utility-scale electricity genera-tion; localized, small-scale microgeneration; low-energy transportation alternatives; waste and wastewater energy-capture projects; information and behavioral change programs; and national, policy-level projects that, taken together, offer a roadmap for developing nations to strengthen their economies while limiting carbon intensity.

Alternative Power Production

One of the primary objectives of a low-carbon approach is to swap out fuels with high carbon emissions and replace them with fuels having low or no carbon emissions. For example, switching from coal- or oil-fired power plants to those powered by natural gas significantly lowers emissions, as well as particulate levels of SO_x, and NO_x, with the added benefit of improved air quality. Given a primary base load (constant power demand through day and night) served by gas-fired cogeneration, new, utility-scale renewable energy—such as hydro, solar, wind, or geothermal facilities, with capacities of 1 MW to 1,000 MW—can be readily scaled up to achieve reliable energy access.

In countries lacking fossil resources within their borders, the cost of importing coal, oil, or natural gas for electricity generation can drive the cost of electricity to an unaf-fordable \$0.20–\$0.30/kWh (in the United States, the average retail price per kWh is about \$0.10). Although the costs of solar and wind power are rapidly approaching com-petitive levels, the upfront costs for hydro, wind, and solar facilities are still higher than those for conventional power generation—and in developing countries, where power demands in growing cities can be between 100 MW and 1,000 MW, the cost difference can be hundreds of millions of dollars. Moreover, at scales over 100 MW, renewable power requires a very large land area or a site at a significant distance out to sea, both of which can entail damage to ecological habitats that outweigh emissions benefits. To power the world's fast-growing cities with low-carbon sources, electric utilities and power plant developers must find the sweet spot that accommodates the use of renew-able resources and environmentally appropriate siting while attracting sufficient up-front financing for large-scale renewable energy.

For rural areas in developing countries, scenarios involving large, centralized plants and the transmission of power across long distances are likely to be more expensive than a smaller-scale, distributed, localized approach. Moreover, even if rural areas do obtain grid access, subsistence farming communities might not be able to afford the utility bills. Here is where small-scale renewable energy makes the most sense. Rooftop solar, for example, at the scale of a single home, can provide enough power for evening lighting, cell phone charging, and small household appliances. Small solar systems that cost less than $1,000 can completely transform the lives of the rural poor.

In contrast to rural areas, high-density urban areas with mixed land uses and power loads offer many more opportunities for advanced energy generation and conversion technologies. For example, cogeneration—the use of waste heat from fossil fuel power plants—can reduce carbon emissions by as much as 30 percent per unit of output.

Generating Power at the Point of Use: The South African iShack

Although almost two-thirds of South Africa's population live in urban areas, a full quarter of city dwellers reside in informal settlements that lack basic infrastructure services. And despite a 1994 government initiative that promised every citizen a brick-and-mortar home, demand for state-sponsored housing has far outstripped government capacity: a backlog of 600,000 units remains. Meanwhile, an estimated 54 percent of those in informal settlements are unable to access the national grid.[8]

The iShack—an eco-friendly structure, complete with solar panels—was developed to address both housing and infrastructure shortages (figure 4.2). The project's founders, a multidisciplinary team from the University of Stellenbosch, have created prototypes of the 40 m^2 (430.5 ft^2) "improved shacks" in Enkanini, an informal settlement on the outskirts of Stellenbosch. Equipped with rooftop photovoltaic solar panels, each shack has the capacity to provide its residents with 25 W of electricity—enough to power three lights, a cell phone charger, and a motion-sensitive exterior alarm.[9]

Ecological design elements include flooring built from locally sourced bricks, a sloped roof for rainwater collection, and a back wall constructed from clay, mud, straw, and hay—a cost-cutting measure that also regulates temperature.[10] Although currently all iShacks are new constructions, the research team envisions retrofits in the future, as well as designs with enough power to charge larger appliances such as radios, televisions, DVD players, and refrigerators. Every new iteration reflects the premise that eco-friendly technologies can be used to simultaneously upgrade settlements and boost local entrepreneurial capacity.

Although more expensive than a regular shack—the iShack costs about Rs 5,600 ($660), whereas the cost of a regular shack can range from pennies (when built with recycled scrap metal) to Rs 5,000 ($582—the average price on the burgeoning informal real estate market), the iShack offers value that extends far beyond its role as a power source.[11] The iShack's pay-as-you-go system creates employment opportunities for the local community; the Stellenbosch team is training a group of community members as "iShack agents"—staff who will be responsible for installing, reading, and repairing outdoor digital meters as well as for collecting monthly payments.[12]

Figure 4.2
iShack being installed in an informal settlement in Enkanni on the outskirts of Cape Town, South Africa. Photo Credit: Sustainability Institute Innovation Lab.

The project's founders anticipate that in addition to boosting local employment, the new power source will foster economic activity. With reliable energy, residents will be able to launch informal income-generating enterprises such as cooking or welding, and children will be able to study long into the evening. Moreover, the improved dwellings will give residents more of a stake in their future, motivating them to fight for legal tenure—which can build social capital, improve community stability, and reduce crime in the notoriously unsafe shantytowns. With the help of a $250,000 grant from the Bill and Melinda Gates Foundation, up to 100 shacks will be constructed to help lay the groundwork for future scale-ups. If the prototypes are successful, the project will stand as an exemplar of the desirability of eco-friendly, bottom-up development.

As a potentially sustainable solution to long-standing social problems, the iShack demonstrates that integrated design can occur even in the context of self-built construction and informal settlements. Combining housing, local materials, renewable energy, and community empowerment, the structures bring families lighting, communication, entertainment, and security through a low-carbon approach that offers economic benefits—including motivation for tenure acquisition.

Microhydro in Nepal

With a per capita income of $340 and 40 percent of its population living below the poverty line, Nepal is one of the world's poorest countries. As of 2014, 84 percent of Nepalese still lived in remote rural areas, where only 15 percent of households have access to electricity.[13] Nepal's mountainous terrain makes grid connections to remote villages prohibitively expensive. Lacking other energy sources, residents burn wood for heating, cooking, and lighting—in fact, wood, crop residues and animal dung combustion accounts for 86 percent of the country's total energy consumption.[14]

Relying on woody biomass as the dominant energy source is highly unsustainable. As energy consumption continues to grow, peak demand is increasing by 10 percent annually. Burdened by an inadequate national grid (with notorious 16 hour blackouts), even urban citizens resort to labor-intensive wood harvesting, despite the human and environmental threats posed by the resultant deforestation.[15]

Although Nepal lacks indigenous supplies of gas, coal, or oil, the landlocked mountain nation does possess one overlooked resource: its seemingly ubiquitous, downward-flowing streams. Recent advances in microhydropower generation hold the potential for Nepal to overcome the compounded issues of poverty and energy deficiency. Hydropower is both ample—the country has barely exploited 2 percent of its potential—and relatively uncomplicated in that local residents can be trained to operate and manage the microtechnology. As a result, national and international development organizations have begun to spearhead partnerships to expand microhydro capacity.

The efforts began in earnest in 1996, with the UN-sponsored Rural Energy Development Program. Originating as a pilot initiative in five remote communities, the project was extended to 10 districts in 1998 and to 15 more in 2000. The program also formed the basis for the National Hydropower Development Policy of 2001, which later became the foundation for Nepal's 2006 National Rural Energy Policy and subsequent five-year plans.[16]

The genius of this innovative program lies in its relative simplicity. Whereas large-scale hydropower projects are costly and often disruptive to downriver communities, microhydro is inexpensive and has minimal environmental impact. Setting up the technology requires only minor rechanneling of rivers and the installation of 10 kW to 100 kW hydroturbines. These relatively small-scale turbines are driven by rushing water that is fed back into the river just a few hundred meters downstream, providing enough energy to power an entire village while avoiding flow disruption and major ecological impacts.

The financial costs of minihydropower are minor compared with the cost of expanding the national grid. Whereas the government has implemented conventional hydropower schemes with an estimated average cost of $2,800/kW, private hydropower developers have built significantly less expensive systems, with costs as low as $1,000/kW.[17] To finance these projects, the Nepal Electricity Authority committed to providing up to 80 percent of the funding required for capital investments, leaving communities responsible for securing the remainder.[18] Additional funding was obtained through matching grants from a variety of sources, including the UNDP, the World Bank, NGOs, and local banks. These grants made it possible to gradually shift from public to community funding—a shift that was compelled, in large part, by programmatic mandates that required villages to finance and help manage the systems.[19]

When a village decides to undertake a microhydro project, residents undergo a six-month orientation and training program focused on participatory development and the benefits of rural energy systems. To increase women's engagement in decision making (Nepali society is characterized by long-standing gender bias), residents then form men's and women's community organizations. Once the training is complete, community organizations take responsibility for all aspects of hydropower creation—from digging diversion channels to stringing power lines and installing lights. In addition to creating jobs and strengthening community buy-in, villagewide participation enables local residents to become skilled in maintenance and repair and in transmitting technical know-how.

Since 1996, the program has enabled nearly 60,000 households in Nepal's poorest rural communities to access reliable and sustainable energy. Equally important are the economic outcomes—whereas rural residents had previously devoted several hours each day to procuring firewood and water, they now have time for income-generating activities such as welding, baking, planting vegetables, agricultural processing, and studying. As of 2009, UNDP estimated that the projects had led to an 8 percent increase in rural household incomes, reduction of annual household energy spending from $41 to $19, and the creation of 40 new businesses for every new microhydro station constructed.[20] In addition to generating income, the new energy supply has boosted enrollment in primary education, empowered women, improved child and maternal health, and enhanced water quality and access to sanitation. As installations continue, the United Nations projects that 15 percent of Nepal's energy could ultimately be generated by microhydropower.

A Disruptive Innovation: Power Plants for India's Villages

Solar energy holds great promise for those beyond the reach of transmission and distribution lines—and even for those who are technically "wired" but receive no power. This is the case in many districts in India, where the companies that supply electricity to large urban areas, including industrial facilities, cannot spare the power needed to serve remote villages.[21]

For the 237 million Indian citizens with no access to power,[22] midscale distributed PV installations offer hope, but only if the market can support the investments. Although the nation's regulatory framework fully supports rural electrification, many in the renewable energy business have regarded the installation of PV power plants in remote areas as a financially risky proposition.[23] The demand is certainly there; concerns have focused on the ability and willingness of the poor to pay—in particular, to pay enough to ensure a steady revenue stream. In addition, many localities have had negative experiences with substandard PV products and technology.[24] Finally, rural electrification is more expensive than its urban counterpart. In India, rural consumers generally finance more than 20 percent of the capital investment in renewable power generation.[25]

In Uttar Pradesh, India's most populous state (population 200 million), Omnigrid Micropower Co., Pvt., Ltd. (OMC) saw market potential for solar-powered generation. In 2012, OMC developed a unique business plan that leveraged demand from the telecommunications industry, thereby guaranteeing a workable bottom line for ten years while providing power to some of India's poorest citizens.

As of 2013, India had about 400,000 cell phone tower base stations—over 150,000 of which were off-grid, powered by dirty diesel generators. Since 2012, tower company operators have been under a government mandate to transition 50 percent of their rural towers to renewable power sources by 2015.[26] Renewably powered base stations are not an inexpensive proposition, but with surging cell phone use (India has more than 960 million subscribers, second only to China),[27] the telecom industry has a strong interest in weaning cell towers from diesel fuel, which is much more costly than grid-connected energy.

OMC's insight was that by taking advantage of the anchor demand—and guaranteed revenue stream—from telecom towers, it could make remote electricity "bankable."[28] Through its Community Power program, OMC has constructed eleven 9 to 18 kW micro solar power plants, costing about 50 lakh ($75,385) each and sized so that their excess power could be transmitted to local communities via minigrids.[29] OMC's disruptive innovation was to effectively link the needs of two different clients. By adroitly amalgamating the telecom industry's power demands with those of the poor, OMC created a viable combination (see figure 4.3).

The photovoltaic micropower plants come without the high capital costs and long construction times of coal-fired plants (three months vs. five to seven years) and eliminate the energy losses associated with power transmission. The power plants can be supplemented by wind energy or by generators running on biogas. The plants are

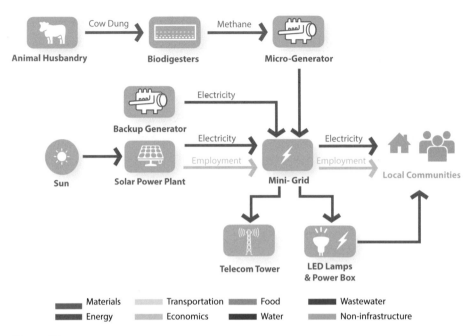

Figure 4.3
Diagram of OMC power plant and minigrid, Uttar Pradesh, India. Credit: Hillary Brown and
Logman Arja.

situated near the tower base stations, with power cables connecting the two.[30] Local
agents supply the nearby communities with minigrid-adapted power equipment. In
addition to leapfrogging the grid, the arrangement also eliminates expensive wiring
for household service. OMC leases out portable lanterns powered by the minigrid,
which take advantage of the effectiveness (and extraordinary efficiency) of LED lighting technology.

Householders in the villages served by OMC found that they could afford the company's basic illumination service, which costs about Rs 100 ($1.50) each month, versus
Rs 180 ($2.70) for kerosene.[31] Each morning, "light wallahs" pick up the lanterns for
recharging and return them to the customers later in the day (see figure 4.4).[32] Households that can afford to can also rent a charged battery unit, which can power devices
such as fans and TVs. All services are prepaid.[33] As of 2013, OMC was diversifying
its products, leasing small refrigerators and high-efficiency irrigation pumps—and, for
those with tablets, streaming live TV via Wi-Fi.[34]

What is game changing here is the unique nature of the partnerships. OMC's understanding of the telecom industry's needs enabled it to provide communities with power.

Figure 4.4
OMC's light wallahs carrying recharged lanterns, Uttar Pradesh, India. Photo Credit: Pär Almqvist, courtesy of OMC Power.

Most interesting is that over half of OMC's revenue comes from residents—refuting the notion that the poor cannot be the source of a bankable revenue stream.[35] OMC has also found another advantage to the arrangement in that whereas full payback for similar infrastructure could typically take six to seven years, OMC breaks even within six months.[36] The telecom tower companies are pleased because not only has the cost of operating the towers dropped significantly, but rural cell phone business has surged.[37] Finally, the communities have realized an important benefit: direct job creation. OMC employs between 10 and 15 workers at each tower.

Today, an artisanal textile shop in the small village of Meer Naga, once sporadically electrified, has secure lighting.[38] Workers can double their incomes by working after dark, children can study, and women can cook after sundown, without the cost and pollution of kerosene. The Community Power program offers huge opportunities for expansion; OMC plans to scale up from tens to hundreds of plants, mostly in Uttar Pradesh and the northeastern states. At the World Economic Forum in 2014, OMC won an award for profitably delivering clean energy to the poor.[39]

Hybrid Pumped Storage Systems

Although many emerging economies are blessed with ready—even excess—sources of wind and solar energy, supply and demand are not always in sync. Thus, the successful integration of large-scale, intermittent energy sources into an electrical power grid depends on storage. Storage is essential to ensuring reliability and stability (frequency control). Indeed, with use of renewables on the rise, it is storage capacity that will largely govern the pace of the transition from carbon-based to alternative energy.

Pumped storage hydro (PSH) is a cost-effective and conceptually simple technology traditionally used for load balancing: evening out energy peaks. It is a closed-loop system in which off-peak electricity is used to lift water from a lower reservoir to a higher one, thereby storing it as potential gravitational (hydraulic) energy, capable of generating power on demand. Typically, the reservoirs are sited to take advantage of natural topography (see figure 4.5).

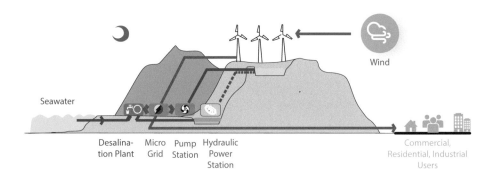

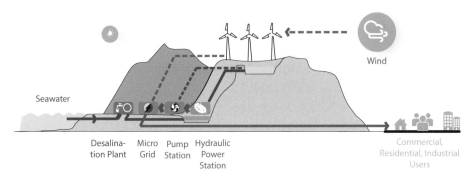

Figure 4.5
Diagrammatic section (nocturnal and diurnal) of pumped storage system and grid, El Hierro, Canary Islands, Spain. Credit: Logman Arja.

For large-scale energy storage with an efficiency range of 75 percent to 85 percent, PSH—which is essentially a giant battery—is the technology of choice throughout the world,[40] global PSH capacity being some 127,000 MW.[41] Although most installations are in Europe, Japan, and the United States, rapidly growing economies such as China and India often use PSH to integrate storage capacity into new or existing hydro dams.

But what if PSH were coupled with power-generating wind farms? That was the question posed to Endesa, a Spanish energy company, by the Canary Islands Institute of Technology and the elected officials of El Hierro, the smallest island in an archipelago off the northwest coast of Africa. At the time, El Hierro's sole power source was costly diesel fuel, imported by barge. Today, through the use of off-the-shelf technology, the island (population 11,000) is powered entirely by renewable energy.

Thanks to the Atlantic trade winds that gust across its steep cliffs, the island's five new 2.3 MW wind turbines produce more energy than is required during peak demand, creating a perfect opportunity for PSH. Powered by the surplus energy from the wind turbines, especially at night, pumps move water 710 m (approximately 0.5 mi) uphill—from a lake constructed near sea level to a second reservoir, which was formed by modifying an existing volcanic crater.[42] When the wind dies down, water is released to four hydroelectric turbines that supplant the wind power and even out the flow of electricity. This unique hybrid system provides a constant power supply that satisfies the island's 47.4 GWh annual demand, including its peak demand (7.6 MW).[43] The former diesel generation system remains in place in a backup capacity (see figure 4.6).

Because it does not rely on huge reservoirs capable of supplying power continuously, this combination of wind and water is environmentally superior to a conventional hydroelectric dam. In a PSH system, water is recycled through pipelines between the reservoirs, and only a small amount is required to top up whatever is lost from leakage and evaporation. By jettisoning diesel power, El Hierro was able to eliminate the annual cost of 40,000 barrels of oil and lower its annual carbon footprint by 18,700 metric tons (20,613 tons).[44] The system had to pass rigorous environmental criteria, which the developers met by removing and replanting native shrub habitat and protecting existing cypress.[45]

Notably, the project was conceived and implemented through multiparty agreements. The Spanish government and the European Union were involved (the latter providing a grant of €35 million); other actors—and eventual owners—included the Canary Islands Institute of Technology (10 percent ownership), the Council of El Hierro (60 percent), and Endesa (30 percent). A new public-private enterprise, Gorona del Viento El Hierro S.A., manages the system.[46] The total cost of the initiative came

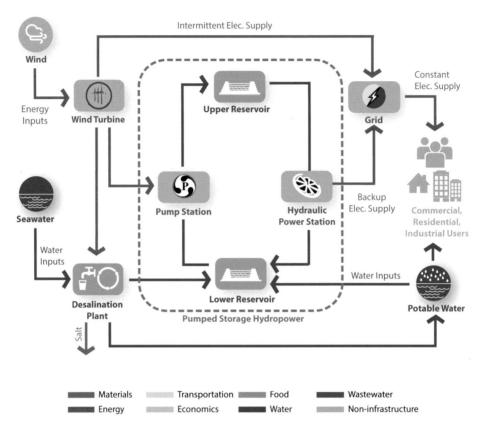

Figure 4.6
Diagram of El Hierro pumped storage system and grid, El Hierro, Canary Islands, Spain. Credit: Logman Arja.

to approximately €90 million ($113 million); given the generous backing from grants, costs may be recouped within ten years.[47]

A collateral effect of this ambitious undertaking was newfound pride in the island's power autonomy, which helped foster another low-carbon innovation when the El Hierro government set a goal of replacing all the island's cars with electric vehicles by 2020.[48] These vehicles are to be integrated into the hydro-wind system to use (as well as store) excess wind energy, yielding a dynamic, unified power generation and storage array.

The wind and water project, which came online in June 2015, demonstrates that isolated islands can rely on local renewable resources to meet their energy needs. Since there are numerous similarly sequestered islands across the globe, El Hierro affords a superb template for adaptation. As of 2015, a similar installation was under construction on Greece's mountainous Ikaria Island.[49]

Even in mainland locales, hybrid wind-and-water systems offer promise. For rural areas in developing nations that have solar and wind potential, renewables—coupled with storage systems like PSH—may make it possible to leapfrog conventional infrastructure. In fact, in remote areas, distributed generation may be more economically feasible than grid expansion.

Although PSH requires two reservoirs, separated by a minimum elevation of 300 m (about 1,000 ft) and at a distance of no more than three km (about 0.3 mi),[50] such systems can be feasible even in the absence of naturally suitable topography. In August 2014, the U.S. Federal Energy Regulatory Commission issued a preliminary permit for Green Energy Storage Corporation to study the Sacaton Pumped Storage Project in the flat, arid landscape northwest of Tucson, Arizona. Currently occupied by a decommissioned open-pit mine, the site could be transformed into a 100MW solar plant and a 150 MW PSH facility. In the proposed arrangement, solar energy would be used to pump reservoir water, held in the depth of the pit, to an upper reservoir located on what was once the dumping ground for waste rock. As the water drops back into the pit, hydroelectric turbines will churn out reliable power for the grid.[51]

Some of the most vigorous winds that buffet Haiti are to be found in the northwestern part of the country, not far from the steep changes in elevation required for PSH. According to Worldwatch Institute researchers, not only is Haiti's geology well suited to hybrid wind-and-water investments, but such systems could also be used to supply renewable power to energy-intensive desalination plants,[52] which would be a boon to Haiti, a water-stressed country where 3.59 million people—40 percent of the population—lack access to clean water.[53] Finally, the wind-powered PSH system could address some of the fragility of Haiti's grid. Since much of the nation's grid power comes from its hydro dams, generating capacity decreases from 90 MW during the rainy season to 45 MW during the dry season.

Despite the promise held out by renewably powered hybrid systems, the barriers for Haiti and many other developing nations are significant. The principal difficulty is obtaining the funds necessary to install such systems in remote and mountainous regions lacking road infrastructure.[54] One option for meeting such needs is a multilateral, public-private enterprise along the lines of that used in El Hierro.

For many small islands, the high cost of importing fuel hinders development—in part by limiting economic activity to agriculture, which is highly sensitive to market dynamics. Reliance on local resources, such as hybrid wind and water, can not only improve energy quality and energy security but also clear a path for social and economic advancement.

Alternative Heat Production

In emerging economies, energy from the sun and from biomass is proving to be an ideal alternative to fossil fuel. One example is the rooftop solar water heater—a simple, relatively inexpensive technology that harnesses the sun's energy for domestic hot water. Brazil, India, and China are currently the leaders in the use of solar water heaters, but there is ample opportunity for other developing nations to follow their lead, thereby reducing the need for new coal-fired power plants. In China alone, for example, 27 million domestic solar units generate heat that is equivalent to the output of 49 coal-fired plants.[55]

Biomass combustion is commercially available at scale and is ideal for district heating of buildings in cooler climates. The relative abundance of biomass in rural areas that may lack access to other resources makes it economical and practical. And, provided that the biomass is a residual from another activity (such as sawmilling) and does not compete with other uses (such as food production), it is considered renewable.[56]

Heat from the Sun: Municipal Innovation in Betim, Brazil

Among upper-middle-income countries, Brazil has been a leader in renewable energy development policy. In the second half of the twentieth century, Brazil underwent five decades of intense urbanization and outstripped other Latin American countries in the construction of power generation capacity. And by 1990, up to 49 percent of its energy came from renewables—predominantly hydropower, followed by solar energy and biofuels.[57]

Although by 2014 that percentage had decreased to about 40 percent as a result of a reduction in its hydropower generation,[58] Brazil has been a world leader in transportation biofuels (ethanol made from bagasse, a sugarcane residue), which can be used to power buses and taxis, as well as automobiles with "flex-fuel" engines (which allow them to run either on gasoline or biofuel).[59] Although new fossil fuel sources (such as the Gasbol Bolivia-to-Brazil gas pipeline) have set back Brazil's overall percentage use of renewables somewhat, the country continues to pursue a diversified approach to decarbonization.

Throughout Brazil, practically all domestic hot water has been produced by in-home electric heaters—causing sizable spikes in the evenings, when most of the population showers.[60] Betim (population 440,000)—a large industrial city in the state of Minas Gerais in southeastern Brazil—became a member of ICLEI's Cities for Climate Protection campaign in 2002. On the basis of a greenhouse gas inventory undertaken in 2002, the local government saw an opportunity to reduce emissions.[61] Forming a partnership with Cobhab (the Minas Geras Housing Department) and GEMIG (the state electrical utility), the city devised a plan that would deploy solar thermal energy to offset peak power demand while making electricity more affordable for the residents of the city's public housing estates. Between 2004 and 2007, GEMIG—at its own expense and without any cost to households—installed 1,356 solar water heaters (SWHs) on four housing estates.[62]

In 2009, the same partnership undertook a community survey to evaluate the effectiveness of the program. The responses to the heaters were positive, and the program's success strengthened residents' commitment to the use of renewables. Typically, the readings showed a 25 percent savings in energy consumption per family, with average consumption dropping from 95.11 kWh/month to 76.01 kWh/month. Also, because the heaters substantially reduced monthly bills, residents became eligible for tax exemptions awarded to homes with low energy consumption, increasing energy savings per family to 57 percent.[63]

In 2008, thanks to the program's success, Betim instituted a requirement for solar water heating for all residential buildings with more than 30 units.[64] The program has also had a multiplier effect in other public and private housing. Today, Brazil's largest city, São Paulo, requires solar hot water for all new buildings larger than 800 m² (8,611 ft²).[65]

Yet another outgrowth of the program was the establishment, in 2007, of the Reference Center in Renewable Energy, a Betim-based organization whose mission is to promote GHG reductions, raise awareness of affordable renewables, engage in outreach to other communities, conduct technical training and workshops in renewables, and create jobs related to energy efficiency. The Reference Center has also helped to replace inefficient street lighting and to advance local ethanol production.[66]

Today, mandates for incorporating solar hot water in new construction represent a strong worldwide trend at both the national and local levels. Under India's energy conservation codes, for example, large commercial, centralized hot water systems and residential development that meets a certain square meter threshold must obtain 20 percent of their energy from solar. Uruguay also mandates solar hot water for many commercial building types, and China plans to require it nationwide in certain types of construction.[67]

Village District Heating in a First Nation Settlement:
Oujé-Bougoumou, Quebec

The use of renewables to heat buildings has hardly kept pace with the deployment of solar electricity or solar heating; nevertheless, it may be economically attractive in some areas. For decades, local mining and forestry interests in James Bay—a large, heavily forested territory in the northern part of Quebec—have severely threatened the traditional inhabitants' subsistence economies and culture.[68] Last relocated in 1970, and living in scattered and impoverished conditions in a subarctic climate, the members of the Cree First Nation fought for title to a portion of their original lands. Finally, in 1992, Canada ceded jurisdiction and agreed to contribute financially to the construction of a permanent new village, Oujé-Bougoumou, which was to be designed by Douglas Cardinal, an Aboriginal architect.[69] Relying on approaches that are in keeping with the philosophy and traditional practices of the Cree, the village and its energy system have proven that "energy can be generated as if communities and the environment mattered."[70]

In accordance with the credo to "harvest only what is required for use, and find a use for all parts of any items harvested," the Cree looked locally for appropriate and sustainable resources.[71] Given its latitude (49.5 degrees north), the village had little or no ability to use solar energy. Instead, Oujé-Bougoumou leaders elected to create the first biomass-fueled district heating system in rural Canada, modeling it on similar systems used in Europe—particularly in Scandinavia.

The system's fuel is "harvested" from the waste wood and sawdust of local sawmills—refuse that the logging industry previously had simply burned.[72] Despite higher initial costs, but with the prospect of lower costs in the future, the village built a central plant consisting of a 1.4 MW wood-chip boiler, which is backed up by traditional oil-based heating. Fueled by the sawmills' refuse, which is purchased for just $6 per metric ton ($6.21 per ton), the 1.2 km (0.75 mi) network of underground pipes supplies 20 public buildings and 140 homes with space heating and hot water. A fully automated boiler system controls every part of the process, from fuel infeed to ash handling.[73] To get the most benefit from the system, all buildings and homes in the village were designed to exceed Canada's highest level of energy efficiency (known as R-2000).[74] In 1995, when the system began operating, biomass provided 89 percent of the community's energy but accounted for only 19 percent of fuel costs.[75]

Although the Canadian government and Hydro Quebec made small contributions to the project, the system was largely self-financed, at a cost of C$46 million (US$42.5 million); residents pay a fixed percentage of their income into a fund that covers capital and operating costs. In recognition of its leadership—and for achieving annual reductions of 2,300 metric tons (2,535 tons) of CO_2, 3.8 metric tons (4.2 tons) of SOx, and 1 metric ton (1.1 tons) of NOx, Oujé-Bougoumou has received numerous international awards and was invited to exhibit the project at the World's Fair Expo2000 in Hannover, Germany.

The benefits of the system are manifold. To begin with, for each thermal megawatt produced from biomass, the cost is 8 percent of what it would be to produce the equivalent amount of energy from oil, and 3.4 percent of what it would be to obtain the energy from grid-based electricity.[76] The savings accrued by using low-cost fuel go toward a fund designated for the construction of additional homes and other cooperatively owned projects. All dollars remain within the community. According to Canada's Department of Natural Resources, the Oujé-Bougoumou system is highly replicable for the large number of remote Aboriginal communities that wish to practice self-sufficiency and develop in ways that reflect their traditional beliefs.

Managing Waste for Energy

The recovery of biological waste and the capture of methane emissions from landfills and wastewater treatment plants can play major roles in fostering a low-carbon future, particularly in urban areas. Such initiatives not only reduce the harmful emissions associated with methane but also repurpose biogas as a replacement for fossil fuels. In developing countries, low-carbon growth will depend on such closed-loop, ecologically sound recovery approaches to water, wastewater, and solid waste networks.

Generating Energy from Wastewater in Burundi

In the tiny village of Kigutu, Burundi, a thriving medical facility is spurring healthy, sustainable growth. Building on the success that it has achieved since its founding in 2006, a nonprofit called Village Health Works (VHW) has planned a new, 4,600s m^2 (50,000 ft^2) regional surgical hospital to serve pregnant women, particularly those at risk of birth complications. Because the new hospital will more than triple the energy, water, waste, transportation, and communication demands of the campus, it became the impetus for rethinking the facility's infrastructure systems.

VHW subscribes to a holistic vision of care that focuses not just on symptoms but also on the root causes of poor health. Thus, the infrastructure plan was developed through an interdisciplinary process, with the goal of designing infrastructure systems that would improve the environment, human health, and economic well-being in far-reaching and meaningful ways. The resulting plans for energy, water, solid waste, sanitation, and road networks were based on a set of core principles that represent a significant departure from the Western models often implemented in Africa. Designed to be robust and reliable, and to meet international standards for best practices, the facilities also incorporated local materials, optimized the use (and reuse) of energy and materials, and reflected local construction and maintenance capacity.

In the course of the planning process, the local community identified a number of concerns:

• Deforestation, which stemmed from the use of wood as the primary cooking fuel, and whose effects included landslides and the erosion of steep slopes
• Malnutrition, which resulted from inefficient farming techniques
• Respiratory illnesses caused by exposure to indoor cooking fires
• Extreme poverty, which was linked to limited job opportunities and technical training

The wastewater and solid waste treatment schemes are central to the plan. Waste streams from the wastewater treatment process (primary solids), agriculture (cow manure), and cooking facilities (food scraps) will be fed to an anaerobic digester that produces biogas and biosolids (digestate). The biogas will be used as cooking fuel for food preparation, providing sufficient energy to serve the entire VHW hospital and hospital staff—nearly 250 people. Meanwhile, the biosolids will be combined with other biodegradable waste in a windrow composting facility, and the resulting compost will be applied to agricultural fields to increase crop yields. Plans also include specific recommendations for improving roads, electricity supply, drainage, and drinking water access (figure 4.7).

The waste collection arrangement has the potential to address a number of community concerns. Using biogas for cooking fuel will reduce the need for the wood-based

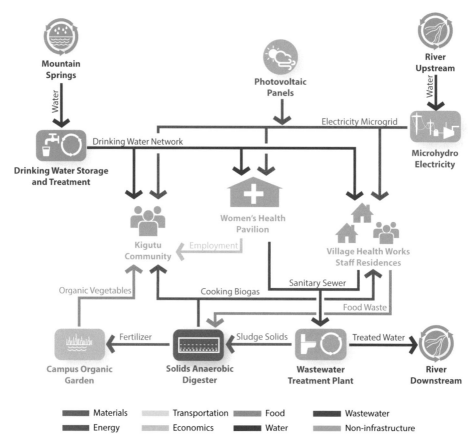

Figure 4.7
Village health works integrated infrastructure, Kigutu, Burundi. Credit: April Schneider.

fuels, mitigating deforestation.[77] (In Burundi, which has a deforestation rate of nearly 10 percent per year, the application of such technology has the potential for widespread impact.)[78] Other benefits of biogas include better health (from improved indoor air quality) as well as significant reductions in the time spent cooking and collecting wood. Taken together, these benefits will increase participation in income-generating endeavors. In addition, digestate enhances composting processes, producing fertilizer that will increase crop yields and thereby combat malnutrition. Finally, the plan assigns high priority to training, capacity building, and continuity planning, not only to ensure proper upkeep of the infrastructure for years to come but to create a network of skilled laborers that can be expanded throughout the country. The VHW project demonstrates the potential, through integrated planning and thoughtful implementation, to facilitate synergies among systems; address health, economic development, and environmental stewardship; and enable physical infrastructure to have wide-reaching impacts.

Kilowatts from Cow Patties: A Biogas Breakthrough in Ibadan, Nigeria

Rapidly urbanizing centers face the compound challenges of energy poverty, poor sanitation, and escalating solid waste generation, conditions that impair public health and undermine economic development. Through innovative, crosscutting projects, communities are finding cost-effective, environmentally sound solutions to these problems.

Of the solid waste produced by low- and middle-income urban centers, more than 50 percent is typically organic and therefore biodegradable.[79] Biogas-producing waste facilities—anaerobic biodigesters—can accommodate multiple waste streams, including not only domestic organic waste but also human, sanitary, animal, and agricultural waste.[80] In addition to producing a low-carbon fuel (biogas), such systems reduce or obviate the need to grow crops for biofuel.[81] Moreover, the residual slurry can be substituted for synthetic fertilizer, helping restore soil quality in agricultural areas that have been intensively farmed. The slurry has the added advantages that it can sometimes act as an alternative pesticide or fungicide and that it has shown value as food for fish production.[82]

Biodigesters function at a wide range of scales. Domestic brick, concrete, and polyethylene units have been adopted widely in China (8 million) and India (4.5 million),[83] and they have also made incursions into Central and South America, where development agencies have successfully marketed them as affordable sources of both energy and fertilizer. Thanks to the myriad small- to medium-scale digesters in use—predominantly in rural areas—a great deal of practical knowledge about the uses and limits of this basic technology has been documented and published. Anaerobic biodigestion is only beginning to make headway, however, in developing economies at larger, municipal scales.

Although commonplace in industrialized countries, the sophisticated and automated anaerobic digestion technology needed to support midsize and large biogas plants has not yet been widely transferred to emerging economies. Technology transfer will require investors, knowledgeable entrepreneurs, and willing local authorities to sponsor the scaling up of digesters for urbanizing areas. It will also require the skills to support the construction and operation of large plants, as well as user education and changes in waste-handling behavior.

Ibadan, one of Nigeria's largest cities (population three million), has experienced an influx of refugees fleeing political unrest, and the resulting unplanned urbanization has led to intensifying environmental problems. The municipality, which sprawls over an area of 3,080 km^2 (1,190 mi^2), lacks adequate energy, water, sanitation, and waste collection infrastructure. Moreover, the Bodija Municipal Abattoir, a statewide slaughterhouse handling over 1,000 head of cattle daily, empties unprocessed wastes into open drains, carrying high levels of diseasing-causing microorganisms and polluting the groundwater and river water used for domestic supply.[84]

In 2008, working with several institutional partners—the Nigerian branch of the Global Network for Environment and Economic Development Research; the University of Technology in Thonburi, Thailand; and UN Habitat's Sustainable Ibadan Project—a local civil engineer and founder of a new NGO installed a flagship 3,000 m^3 (712,596 gal) capacity bioreactor designed to capture 1,800 m^3 (2,354 yd^3) of biomethane per day.[85] With a design adapted from the anaerobic reactors used to treat agroindustrial waste in Thailand, the plant produces 60 percent to 70 percent methane (the rest is CO$_2$, along with other impurities), which can be used as cooking fuel, generating about 0.5 MW of cooking gas. The system produced a return on the $0.5 million investment after three years.[86]

The biodigester, an airtight reservoir, receives the feedstock—in this case, the abattoir waste. In this oxygen-free environment, acidogenic bacteria (fermentative microbes) digest the biomass into fatty and volatile acetic acids, which are then metabolized by methanogens (methane-producing microbes) to produce a methane-intense gas and a phosphorus- and nitrogen-laden effluent—a slurry that can be used as fertilizer. The digestion process takes from one to more than three weeks to complete.[87]

In addition to abating water pollution, the plant mitigates about 0.2 metric tons (0.22 tons) of carbon dioxide annually and provides an invaluable, clean-burning fuel that improves public health by offsetting the use of coal, wood, and dung. Finally, it produces 1,500 l of organic fertilizer daily, which is purchased by the Oyo State Fertilizer Board and sold to urban, low-income farmers at about 5 percent of the cost of standard chemical fertilizers.[88]

Significantly, the plant relies on a technology that is applicable not only to domestic organic waste but also to agroprocessing industries, and this technology could be readily implemented in most developing economies. Still, some drawbacks remain in that food prepared with biogas rather than petroleum- or wood-based fuel requires longer cooking times, and when biogas replaces charcoal, some food may also lose flavor. Perhaps more important, biogas generated from human or animal excrement may be associated with considerable cultural taboos that would interfere with its use in cooking or as fertilizer.

In 2009, the Ibadan project was recognized as a Technology Pioneer by the World Economic Forum; that same year, it was a prizewinner at the Technology Museum of Innovation in San Jose, California. The project carries great potential for replication and rollout.[89] More than two billion people worldwide—including 89 percent of the population of sub-Saharan Africa—still rely on fossil fuel or biomass for cooking and heating.[90] Cleaner, renewable fuels can not only provide energy and recover nutrients for use in agriculture but also reduce indoor air pollution and thereby lower global mortality rates from pulmonary disease.

Ennobling the Role of Landfills: Reducing Greenhouse Gases in Belo Horizonte, Brazil
Today's rising anthropogenic methane emissions are correlated with increases in human population, livestock production, and landfilling; leaks from oil and natural gas recovery are another source. When not properly captured, landfill gas alone leads to the worldwide release of between 19.99 million and 59.99 million metric tons (approximately 22 million to 66 million tons) of methane annually.[91] As early as 1993, the U.S. EPA estimated that global methane emissions from landfills could easily be reduced by about 50 percent by using technologies available at the time. Developing nations, however, continue to face significant barriers to methane recovery—principally because of insufficient familiarity with economically viable recovery practices, poorly functioning energy markets, and unclear legal and regulatory frameworks.[92]

There are some unique exceptions. Even as its energy use climbed and the nation became wealthier, Brazil was a leader in decarbonization and diversification of fuels for electricity production, especially between 1990 and 2010. Methane recovery in Belo Horizonte, Brazil's third-largest city (population 5.487 million), illustrates the potential for closed-loop use of waste as an energy source in swiftly urbanizing settlements.

Since the establishment of the Municipal Committee on Climate Change and Eco-Efficiency in 2006, Belo Horizonte has undertaken major projects for GHG mitigation, including solar hot water heating and a rapid-transit bus program.[93] In 2007, Belo Horizonte officials zeroed in on the city's largest single source of GHG emissions: a landfill site that had closed in 2007 after having plagued nearby residents with odors and heavy truck traffic for 32 years.

At the time the facility shut down, it held 17.4 million m³ (22.75 million yd³) of waste that had accumulated since 1972 and covered a land area of 65 ha (161 acres). The city's goal was to address the odor at the landfill, which continued to be a problem, while adding another option to renewable energy choices. To achieve both ends, the city authorized the construction of methane capture and cleaning infrastructure, both of which would be partially funded through the certification and sale of carbon credits under the Clean Development Mechanism (CDM), a program established through the Kyoto Protocol that had been in operation since 2006. This group of facilities includes:

• The Treatment Center for Solid Waste (Centro de Tratamento de Resíduos Sólidos, CTRS), which was developed and run by an Italian company, Asja Ambiente Italia SpA, came online in 2010. The operation includes a 5 MW capacity energy plant, generating an average of 17,479 MWh annually, which is sold to the grid network operated by CEMIG, the state energy utility.[94] In keeping with one of the pillars of infrastructural ecology—the colocation of other public or civic activities with waste management facilities—CTRS also offers a number of other services (see figure 4.8).
• A composting plant, which processes selected organic waste collected from markets
• A recycling facility for construction waste, which yields synthetic gravel used for street paving
• A "seedling station," which provides plants that are used to support the environmental restoration of the landfill site
• A plant that manages hazardous medical waste
• A tire collection unit, which gathers rubber to be recycled offsite
• A waste transfer station, which has been transporting Belo Horizonte's waste to a landfill in Sabará, a nearby municipality, since 2007
• An environmental education center, which is visited by about 144,000 students annually

In recognition of its innovative solutions to waste management, mitigation of the environmental impacts of the landfill, and reductions of GHG emissions, Belo Horizonte receives 6 percent of the value of the electricity sold. In addition, the municipality receives revenue from the sale of approximately 1.3 million certified emission reduction credits, which are being sold in the international market during a ten-year CDM crediting period.[95]

Investments such as Belo Horizonte's are highly replicable. Biomethane extraction has proven profitable for medium to large landfill sites—typically for cities with populations of 100,000 or more.[96] Success depends, however, on a number of factors,

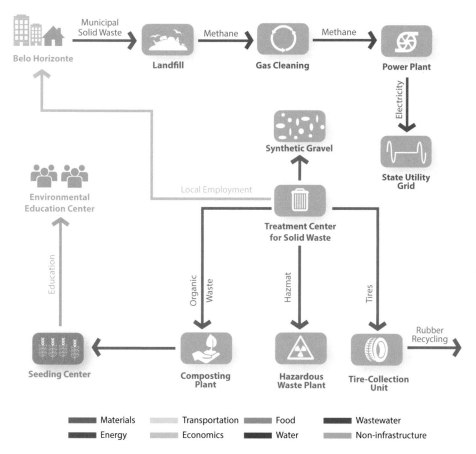

Figure 4.8
Landfill gas to energy and solid waste management, Belo Horizonte, Brazil. Credit: Hillary Brown and Logman Arja.

including the site's climatic and geographical characteristics, the physical and biological attributes of the waste in the area, and landfill design and operation. Under optimal conditions, medium-sized cities worldwide have the potential to reduce their carbon (and methane) footprints, along with their dependence on fossil fuels. And in emerging economies, the coupling of organic waste management with other waste recovery and recycling services offers not only revenue but job creation opportunities, including local employment in construction, maintenance, and operations.[97]

Agroenergy for Ajuricaba: Cross-Sector Successes in Brazil

Itaipu Binacional (IB), the entity behind the world's largest generator of renewable power, finally completed the 8-km (5-mi)-wide Itaipu hydroelectric dam in 1991. In 1994, the American Society of Civil Engineers lauded the project as one of the seven modern Wonders of the World.[98]

Like many other megadams, Itaipu had a controversial beginning. Construction had displaced 59,000 inhabitants of the banks of the Paraná River,[99] eradicated a national park (including Guaíra Falls, a popular tourist site, which was dynamited), and led to land and water degradation and loss of biodiversity. From this perspective, IB's activism over the past two decades—promoting sustainable development, energy access, employment, and environmental conservation—might be seen as partial recompense for its earlier transgressions.

Itaipu Binacional's name reflects its geography: the dam reaps the benefits of a river watershed that divides Paraguay and Brazil. The dam's 14,000 MW installed capacity satisfies 17 percent of Brazil's and 72 percent of Paraguay's energy demand. Its physical assets belong to Eletrobras, one of Brazil's largest power utilities, and ANDE, Paraguay's public utility. The Brazilian government funded all construction costs.[100]

Although about 88 percent of Brazil's electricity comes from hydroelectric generation, in 2002 the country instituted an incentive program to foster the development of additional renewable technologies, including wind, biomass, and small hydro.[101] In December 2009, the Brazilian government set a target of reducing national greenhouse gas emissions by 38.9 percent by 2020. It anticipates that such reductions will come from a combination of renewables, greater efficiencies in buildings and industry, and improved farming practices.

Given that 77 percent of Brazil's population is employed in agriculture—and that agricultural residues are spread widely across the nation's huge territory—distributed power generation from biomass, transformed by biodigesters, offers an elegant means of providing energy access to remote areas.[102] Agroenergy, which transmutes the environmental liabilities of the Brazilian agricultural sector—principally, the methane produced by animal manure, and the chemical fertilizers that pollute watersheds—into electricity and biofertilizer, has become a significant source of additional income for rural communities.[103]

In 2009, the Agroenergy Condominium for Family Agriculture was established in the Ajuricaba River watershed through a partnership between the Itaipu Office of Renewable Energy, the Institute of Technical Assistance and Rural Extension, the Parana State Electricity Company, the International Center for Renewable Energies, and other entities. The cooperative, which includes 33 small-scale family farms, became the focus of one effort on the part of IB to support the production of biogas and biofertilizers. To process the waste from corn production and the farmers' herds (approximately 1,000 head of

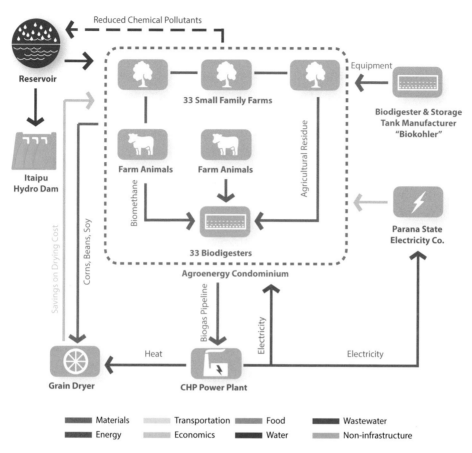

Figure 4.9
Diagram of Itaipu Binacionale's agroenergy condominium for family agriculture, Paraná River basin, Brazil. Credit: Hillary Brown and Logman Arja.

cattle and 3,000 swine), IB installed individual biodigesters at each farm. The farms generate 15,800 m³ (4.2 million gal) of agricultural residue and manure annually—organic waste that yields 266,600 m³ (348,699 gal) of biogas. The biogas is transported through a pipeline 22 km (13.7 mi) long to a centrally located power plant, where it produces heat, electricity, and—after upgrading—biomethane vehicle fuel.[104] Overall, the gas generates 445,000 kWh/yr, sufficient energy to serve about 2,200 area households.[105] The surplus is sold to the state's energy distributor, producing revenue for the farmers. The farmers also make use of the plant's grain dryer, which runs on the thermal energy produced by the plant, to dry products such as corn, beans, and soy, reducing drying costs by as much as 90 percent (see figure 4.9).[106]

In addition to producing energy, micropower generators like those installed at the Agroenergy Condominium eliminate a polluting nuisance, reduce methane emissions, improve farming profitability, and strengthen local development. For example, the Ajuricaba biogas initiative has fostered a small local industry also incubated by IB, Bio-Kohler, which produces the fiberglass biodigesters, storage tanks, and cooking stoves designed for biogas use.[107]

In 2003, having recognized the links between the Itaipu dam's alteration of watershed hydrology, agricultural environmental degradation, environmental harms associated with energy production (in particular, methane emissions from hydro dams), and the region's poverty, IB expanded its vision and mission, stating its intent to "generate quality electric energy, while maintaining social and environmental responsibility, spurring economic, tourism and sustainable development in Brazil and Paraguay." Through a comprehensive social and environmental program known as Cultivando Agua Boa—Cultivating Good Water—IB has built a unique, multidimensional framework for exemplary stewardship involving local communities.

Among the program's achievements are the recovery of 206 microbasins, yielding upgraded water quality and quantity; the restoration of 1,321 km (821 mi) of riverine buffer zones; the remediation of 22,528 ha (55,668 acres) of agricultural lands; the restoration of over 40 million trees; and the creation of two biological sanctuaries and a 13 km (8 mi) biodiversity corridor.[108] Other Cultivating Good Water initiatives involve fostering new practices that improve agriculture in the watershed. These include providing technical assistance for the conversion to organic agriculture, which relies on "green fertilizer" (from biodigestion); introducing no-till farming and the cultivation of medicinal plants; providing nutritional education; and integrating fish farming into existing agricultural activities.[109] Finally, by helping to establish 5 solid waste cooperatives and 25 solid waste associations in the region, the program has improved the lives of solid waste collectors.

In 2004, to further promote regional development, IB established a foundation, the Itaipu Technological Park (ITP). Initially focused on water, energy, and tourism, ITP has become a leading engine of scientific and technological progress in Latin America.[110] Occupying a 116 ha (287 acre) site near the Itaipu dam, ITP partners with several universities to provide training and research grants and to support entrepreneurship in a number of fields, including hydrology, agribusiness, biogas technology, pharmaceuticals, and, significantly, environmental education. In partnership with Itaipu Corporate University—a corporate research arm of IB—ITP also works to improve knowledge sharing across the electric power sector. Finally, ITP's Sustainable Tourism Program is developing a systemic approach to regional tourism, which will add value to the farming and ranching landscape.[111]

Today, IB's laudably sustainable vision extends beyond its customer base. Its enterprising Agroenergy Condominium biogas programs are now being replicated in Uruguay, where Eletrobras, using funding from a group formed by the world's 13 leading energy companies, will sponsor the connection of 31 farms to a central microthermoelectric plant, which will produce 780 m^3 (1,020 yd^3) of biogas daily. (Uruguay is undergoing strong industrial development, particularly in the milk-production industry, which has considerable effects on the environment.) IB has also publicly announced its intention to introduce the technology to countries in Africa and Asia.[112] From energy giant to visionary agent of sustainable development, IB has played a transformative role in the region. In 2015, IB received the Best Water Management Practices award from the United Nations Water for Life program.

Taking Advantage of a Neglected Power Source: Methane Recovery from Hydro Dams

Itaipu Binacional's environmental and social stewardship was initially spurred by its concern for the integrity of its water supply. Other concerns associated with hydropower, however, have only begun to garner attention. According to Brazil's National Institute for Space Research, hydroelectric dams are currently the largest single source of global methane emissions. Collectively, the world's 52,000 largest dams emit 104 million metric tons (114 million tons) of methane each year.[113] Methane is a significant contributor to radiative forcing—and its warming impact, measured over a 20 year period, is 72 times greater than that of carbon dioxide.[114]

According to several studies, methane emissions are particularly high in the large tropical reservoirs built over areas that were once densely forested,[115] where decaying organic matter releases significant amounts of CH$_4$. Meanwhile, new vegetative growth in the drawdown areas (those exposed at low water levels) refreshes the organic input. Where waters are relatively shallow, much of the methane produced at the reservoir bottom is oxidized as it rises to the surface and is transformed into CO$_2$. But in deep reservoirs, such as Itaipu, the release of waters drawn from the *hypolimnion*—the deepest, most methane-rich thermal stratum—contributes substantially to methane emissions. If the methane were to be recovered for use, its electricity generation potential would be substantial. At the Petit Saut reservoir in the Amazon, for example, the generation potential of methane emissions could be between 14 percent and 31 percent of the dam's installed hydroelectric capacity. (The broad range is attributable to natural and seasonal fluctuations in methane emissions.)[116]

When methane-rich waters enter hydroelectric turbines, the dissolved methane is immediately "degassed" (released into the atmosphere). The turbulence of the downstream spillways frees up still more gas.[117] The reservoir's surface, where methane is gently released through its own bubbling action (known as *ebullition*), is yet another source of emissions. Surface emissions have been the hardest to measure, but a study of the Nam Theun 2 reservoir in Laos suggests that methane emissions from ebullition in tropical reservoirs are generally higher in the first few years following the filling of the reservoir, because of the relatively rapid decay of once-living matter.[118]

Several solutions have been suggested for recovering the methane associated with hydro dams. In addition to averting harmful emissions, methane collection would allow the gas to be used locally for thermoelectric generation. A barrier device called a "gate-buoy," for example, could divert the surface waters (which have already been depleted of CH_4) for expulsion through the turbines; properly sizing the barrier would ensure minimal impact on power production.[119] In addition, further upstream in the reservoir, closer to its intake point, bubbling or spraying devices could be used to extract methane from the hypolimnion, degassing the CH_4 into a sealed vessel. According to one study, these various methane collection systems could be cost-effective when the long-term returns from the sale of the renewable power are taken into account, suggesting that methane capture and reuse may be a promising approach for large hydropower reservoirs in tropical areas.[120]

According to the International Energy Agency, world hydroelectric production is expected to grow from 4,000 TWh in 2014 to 4,670 TWh by 2020. Even as dams across Europe and the United States are being decommissioned, investment in hydroelectricity elsewhere—including Brazil, China, India, and Thailand—is on the rise, with support from national development banks as well as private institutional investors.[121] In the face of this investment trend and the associated potential for a surge in methane emissions, it is critical to capitalize on biogenic methane as a means of increasing energy supply in developing nations. Not only will doing so mitigate methane emissions, but also the added power production from this newly mined resource may well reduce the necessity of building additional reservoirs.

Already associated with a number of social and environmental harms—including population displacement, disturbance of fish communities, loss of biodiversity, and damage to ecosystems—hydropower has now been revealed as a contributor to climate change. Given the interdependence of land, water, and energy described here, further investigation of, and investment in, methane capture is called for.

Mining natural sources of methane presents an untapped economic and energy opportunity in many parts of the world. Rwanda is a nation dependent on hydroelectricity for two-thirds of its power production. One of its deepest natural bodies of water, Lake Kivu, holds approximately 60 billion m of methane sourced from volcanic soil and decomposing organics in the lake. The nation's first independent power project, KivuWatt, will reduce the quantities of the pernicious gas, preventing both its harmful atmospheric release and the threat of sudden eruption, which would endanger two million people living along the banks of the lake. The first phase of the 100 MW project, commissioned in 2015,[122] consists of a methane gas extraction facility and a nearby independent power plant. Its final buildout, at an estimated cost of $325 million by 2017, could power 30 villages while radically reducing their reliance on diesel-generated electricity.[123]

Decarbonizing Transportation

As any visitor to Manila will quickly recognize, the intensity of the traffic reflects the strength of the Philippine economy. And in China and India, the growth of car ownership has closely tracked the surge in GDP per capita. Meanwhile, the dramatic increase in car ownership is a major source of growth in carbon emissions per capita. There is no question that private vehicles represent freedom, choice, and a faster, more comfortable journey. But, for each trip, travel by foot, bike, bus, or train reduces carbon emissions by a factor of 2 to 5. So the challenge—for urban planners and transportation planners alike—is to render low-carbon transportation as easy, comfortable, and fast as riding in a car.

In emerging economies, where urban sprawl appears to be the predominant mode of development, decision makers need to foster growth patterns based on low-carbon transportation—not only because they reduce greenhouse gas emissions but because they improve residents' health, well-being, and quality of life. One way to measure the impact of low-carbon transportation networks is to determine the average carbon emissions per person per kilometer traveled. For example, if there are two people in a car instead of one, the same amount of fuel yields twice as many "passenger miles." Scale that up to a bus, and emissions per capita per kilometer really start to drop.

In cities with over 100,000 residents, people cannot walk or bike everywhere they want to go; some form of transportation is needed to carry residents across town, and people need to live close enough to bus stops or train stations to be able to walk or bike to them. These transit nodes, in turn, need to be surrounded by shops, banks, and community facilities, so that they are convenient for everyday trips. Developers must be willing to build, market, lease, and invest in this type of high-density, mixed-use urban environment, and banks must be willing to lend money to the developers. Utilities must be willing and able to construct large-scale, renewable energy power plants, and investors must be willing to support utilities in constructing these new types of power plants. National, state, and local elected officials need to see how all of these pieces fit together to create happier, healthier, low-carbon cities. Finally—and perhaps most importantly—residents must want to live in medium- and high-density, mixed-use neighborhoods and take public transportation to work every day.

In sum, low-carbon transportation requires the long-term vision and commitment of urban planners, transportation planners, developers, utilities, financial institutions, elected officials, and, ultimately, communities themselves. Many models of such an approach already exist, and elements of this vision for low-carbon transportation have been successfully implemented.[124]

An "Environmental City": Da Nang, Viet Nam

Da Nang, Viet Nam, offers an excellent example of comprehensive, low-carbon development and multimodal transportation planning. In 2006, a series of progressive mayors began a citywide effort to establish Da Nang as an "environmental city."[125] Improving air and water quality, traffic, waste management, and ecological networks would directly benefit two of the city's main industries—tourism and fishing; equally important, such improvements would boost health and quality of life for Da Nang's residents.

In the decade since these initiatives began, Da Nang has seen substantial advances, particularly in its transportation network. Though traffic is still intense along the city's major corridors, integrated spatial planning has slowed growth in key metrics—such as vehicle ownership, carbon emissions, and traffic fatalities—in comparison with other cities in Viet Nam.

Although central economic planning has its drawbacks, both transportation planning and land-use planning benefit substantially from the requirement to think about an entire city across a 10- to 20-year time horizon. Nor does private property ownership necessarily support the massive, coordinated investments, undertaken by multiple government agencies, required to implement a vision such as the one animating Da Nang. Da Nang did harness the power of the private sector, however, by presenting a compelling vision of a "green city" that would be safer and healthier, have less traffic, and be more attractive to tourists.[126] Many of the planning concepts and details were developed by multilateral agencies such as the World Bank and the Asian Development Bank. But the overall vision, support, leadership, and implementation stemmed from the strong and visionary leadership of Da Nang's mayors.

Aerial Cable Cars: Linking Latin American Hillside Communities to Downtown

In the past decade, dozens of aerial cable-car transportation lines have been proposed and built in the steep hills of major cities in Latin America. In 2004, Medellin, Colombia, led this effort by constructing the Metro-Cable Line-K, which connects the hillside favelas of Santo Domingo to the city's main subway line (figure 4.10).

The low-carbon transit story in Medellin began much earlier, however. Planning for a high-capacity heavy-rail line was first undertaken in the 1980s, and the transit

Figure 4.10
Medellin, Colombia, Line A cable car connecting hillside neighborhoods with the center of the city. Photo Credit: Carolos Mora, © 123RF.com.

backbone—Line A—opened in 1995, serving the Medellin valley and the city center. Meanwhile, informal settlements continued to burgeon on the steep hills surrounding the valley; without access to the rest of the city, these communities were cut off from jobs and opportunity.

With the construction of Line K, life in the favelas changed dramatically. Cable car stations were colocated with important social services, libraries, and public parks. Cable connections to the subway, rapid-transit bus lines, and local bus networks allowed residents quick, easy transportation to jobs anywhere in the city. Most important, this seemingly simple strategy turned out to be a fundamental driver of social and economic change, accelerating the city's transformation from one of the most dangerous places in the world to one that draws tourists and hosts major global conferences, such as the UN Habitat World Urban Forum in 2014.

Many other Latin American cities with similarly steep terrains have concluded that cable cars have the potential to transform not only their physical environment but also their social structure. They have implemented this idea by using a low-carbon development approach that encourages high-density, mixed-use, transit-oriented urban forms. La Paz, Bolivia; Santiago, Chile; and Caracas, Venezuela, are among the dozens of cities around the world that have found a different model of public transportation than their counterparts in the Global North. Connecting hillside low-income neighborhoods with valley-level central business districts can revitalize urban centers and connect residents to parks and open space outside these areas.

Solecshaw: A Solar Rickshaw in India

Rickshaws have long played an integral role in India's teeming cities. The two- or three-wheeled carts have shuttled residents, visitors, and merchandise along crowded city streets since the 1800s. Despite the carts' storied history, their existence remains contentious. State and local governments have periodically threatened to ban them, citing reasons that range from safety, to air quality, to the need to modernize India's image. Meanwhile, rickshaw proponents note that the carts are the sole source of livelihood for more than eight million of India's urban residents—and that the resulting economic gains are achieved without emissions. In a country where more than 60 percent of GHG emissions come from the transport sector, solar-powered rickshaws can move people and goods without generating carbon.

A recently devised high-tech rickshaw, however, has the potential to settle the debate. The Solecshaw—a portmanteau word derived from "solar," "electric," and "rickshaw"—is a pedal- and solar-powered tricycle that preserves employment while mitigating global warming (figure 4.11). A pilot project spearheaded at the Council of Scientific and Industrial Research (CSIR) national laboratory, the rickshaw is being developed to contend with India's mushrooming urbanization. The vehicle is ergonomically designed to reduce the drudgery and physicality involved in rickshaw driving, allowing drivers to make more trips with less effort. The resulting productivity boost is expected to increase income considerably: drivers of the new carts are projected to earn between two and three times as much as their counterparts.

The main source of the Solecshaw's greater productivity is a 36 V solar-powered, battery-operated motor. An eight-hour charge at a solar photovoltaic station will allow the vehicle to travel up to 40 km (24.8 mi) at a maximum speed of 12.5 kph (7.8 mph).[127] To propel the cart, the driver can choose between pedal operation and motor operation.[128]

In addition to possessing enhanced engineering, the three-seated vehicle comes with aesthetic upgrades, including electric lights, FM radio, and a mobile phone charger. Naturally, these improvements come with a price. Whereas a standard rickshaw costs Rs 8,000–10,000 ($130–$160), the Solecshaw is anticipated to cost Rs 30,000–35,000 ($450–$550), with an expected drop to Rs 20,000–25,000 ($325–$400) when the vehicle becomes mass produced.[129]

Figure 4.11
Example of a type of solar-powered rickshaw being developed in India. Photo Credit: 36Clicks, © 123RF.com.

In light of the project's inherent social mission, developers are formulating ways to facilitate installment financing. CSIR representatives say that they envision loan programs that would allow rickshaw owners to pay the same daily or monthly amount they would pay to purchase a standard rickshaw, but for a longer period; for example, instead of paying Rs 30 for 300 days to own an Rs 9,000 cart, the worker could pay the same daily amount for 700 days to purchase an Rs 21,000 Solecshaw.[130]

Government departments have already shown their support for the Solecshaw. In 2010, the Minister of State for Communications and Information Technology announced a proposal to use the Postal Solecshaw, a solar-powered rickshaw, as a postal delivery vehicle to further the government's efforts to deliver services in an efficient and eco-friendly manner. Numerous other initiatives, both public and private, are being planned to expand the use of the Solecshaws for waste disposal, as transport for disabled people, as food carts, as low-cost taxis for tourists, and to provide mobile health services.

From Used Fat to Fuel: Low-Energy Transport in Volta Redonda, Brazil

In Volta Redonda (population 260,000), an industrial and commercial center in the state of Rio de Janeiro, Brazil, 47 percent of carbon emissions come from the transport sector. In 2005, to address this and other environmental threats, the city launched the Environment of Volta program. In 2007, as a participant in two ICLEI programs—Local Governments for Sustainability and the Cities for Climate Protection Campaign—the government was emboldened to look across multiple infrastructure sectors for prospective mitigation measures. The effort to "solve for pattern" yielded a single inventive solution that simultaneously resolved multiple urban problems.

Fried food makes up a major part of the Brazilian diet; as a result, used cooking oil (oleaginous waste) ends up clogging sewage systems, jeopardizing water quality, and increasing water costs and sewage system maintenance by 45 percent. In Volta Redonda, each liter of oil pollutes 14 times the amount of water that an average citizen uses in a year.[131] Today, under a public-private partnership, used cooking oil is collected and processed to produce about 264,000 l (70,000 gal) of biodiesel fuel annually. The partnership, which was established with the encouragement of the municipal government of Volta Redonda, is between the Cesbra Chemical S.A. Group, the Association of Collectors of Effluents and Solid Waste, and the Banco de Cidadania (Volta Redonda's municipal fund for development, employment generation, income, and shelter), and the initiative is known as the Eco-oil Program.

Producing energy from waste at a low cost of .60 real per l ($2.71 per gal), biodiesel has reduced soot production fourfold (when compared with conventional biodiesel) while lowering sulfur oxide, particulate matter, and carbon dioxide emissions. The Volta Redondo city government eventually hopes to quadruple the amount of recovered oil to a total of 20 percent of vegetable oil consumed.[132]

This simple, effective program returns numerous benefits. First, the reuse of cooking oil averts significant harm from the disposal of oleaginous waste. Second, the sale of biofuel returns financial benefits. Third, Eco-oil builds broad environmental awareness. Of the 130 schools (municipal, state, and private) in the city, 128 serve as collection points for the cooking oil, allowing them to share in the revenue from selling the oil to Cesbra, the company that operates the local oil-processing plant (participating schools and commercial establishments also receive certificates of merit for their involvement).[133] Lastly, the program led to a related commercial venture, which sells the oil's residual solid material as an additive for animal feed.

Low-Carbon Paths Forward

As developing countries experience increasing economic success, their burgeoning middle classes will be able to afford cars, air conditioning, and other energy-consuming appurtenances. And if these consumers follow the same trajectory—with respect to both high-carbon technologies and urban development patterns—as their counterparts in the Global North, global carbon emissions will spike dramatically.

The challenge, then, is to address energy poverty, improve quality of life, and increase economic productivity while keeping carbon contributions level. This chapter explores the vast opportunities for emerging economies to do just that, through strategic investment in alternative energy infrastructures.

One last exemplar is the country of Uruguay. According to that nation's head of climate change policy, Uruguay has significantly reduced its carbon footprint through investments in wind and other renewables. It claims it is on target to reduce its carbon emissions by 88 percent by 2017 compared with the 2009–13 average. Notably, this will be accomplished while avoiding government subsidies or higher energy prices. As of December 2015, renewables provided 94.5 percent of Uruguay's electricity. Uruguay's success may be attributable to several factors. First, it has a diverse renewables mix (hydro, solar, and biomass power) and good natural wind conditions: steady ocean winds and low elevations. Second, wind power is reliable and, importantly, is complementary with hydropower, which is vulnerable to drought, a condition during which wind speeds often pick up. A third factor is an ambitious national policy on renewable energy.[134] Lastly, it is a stable democracy, which assures financial credibility, and its government relies on a competitive bidding environment and strong public-private partnerships with wind developers.[135]

As is evident from the range of examples in this chapter, decarbonization can be achieved at a variety of scales and by using a large number of technologies. Low-carbon development strategies must tackle all stages of the energy cycle: generation, delivery, and use. Careful assessment of environmental assets can reveal potential energy sources or sinks (storage opportunities), which can reduce local dependence on imported fossil fuels. Solutions include the conversion of wind, solar energy, water, geothermal energy, or renewable biomass to electricity, heating, or cooling, and projects can also incorporate innovative means of regulating supply and demand. What is key, in emerging economies, is to devise integrated, synergistic approaches that will attract investment from international funding sources and drive economic growth—without impairing ecosystem services or interfering with essential activities, such as food production.

For developing nations to take the first step toward building a low-carbon infrastructure, public and private gas and electrical utilities must not only work closely together but also be alert to the opportunities inherent in untapped or underused local resources—in particular, waste streams from agriculture, sanitation, domestic cooking, and even municipal solid waste.[136] As always, the goal is to "solve for pattern"—to find the synergistic opportunities that, for example, allow sanitary waste disposal to power electricity generation and nutrient recovery. For most emerging economies in the lower latitudes, domestic hot water from solar thermal energy can be readily implemented, and in higher latitudes, renewably harvested biomass offers strong potential for district heating.

Ultimately, an adaptive, polycentric system is needed to govern the energy sector—at multiple levels and through multiple decision-making bodies.[137] Ideally, such a system would yield a stepped sequence leading from energy poverty to grid connection using short-, medium-, and long-term options.[138] An energy access pathway might begin, for example, with small to midsized biomass gasifiers or minihydro, then move on to minigrids relying on community-based photovoltaics or on a regional wind farm with pumped storage hydro, and finally transition to a regional grid-connected wind or solar farm.

There has been some encouraging progress on decarbonization. Between 2008 and 2013, the renewable energy capacity of 55 developing countries grew by 143 percent to a total of 603 GW,[139] a shift that is attributable to the rapidity with which solar and wind power can be scaled up compared with carbon-based power generation methods. Moreover, settlements that now rely on community-based solar power have found that the higher initial costs level out in time, yielding overall costs that are comparable to those associated with centralized grid power.

The holy grail of low-carbon transportation has two elements: first, a completely walkable and bikeable design; and second, an extensive electric bus and train network powered by utility-scale renewable energy—think Amsterdam, completely powered by wind turbines. A zero-carbon transportation network is not as far out of reach as one might imagine, but planners and transportation ministries in emerging economies will need to carefully resist the temptation to overinvest in road and highway systems, which can undermine sustainable transportation. New low-carbon transport mechanisms, such as aerial tramways, can be overlaid on otherwise impenetrable urban fabrics. And a resurgence of cycling and other carbon-neutral forms of transportation must win out over global replication of the West's dependence on the private car.

Efforts to limit the carbon intensity of the transportation sector will require a range of strategies—from low-cost efforts to encourage walking and cycling to capital-intensive investments in passenger and freight rail or rapid-transit bus lines. The success of such strategies will be bolstered by the fact that in many developing nations, public and nonmotorized transport still dominate.

5

Climate-Adaptive Infrastructure: Responding to Changing Conditions

Climate change . . . represents one of the principal challenges facing humanity in our day. Its worst impact will probably be felt by developing countries in coming decades.
—Pope Francis, "Laudato si," May 2015

On May 24, 2015, the Vatican aligned itself with the scientific community by making an unequivocal statement on the impacts of climate change—focusing in particular on the risk that climate perturbations will reverse recent gains and limit future advances in the eradication of poverty. Although emerging economies have borne little responsibility for the greenhouse gas emissions that have triggered climate instability, they stand to suffer disproportionately from the resulting burdens.[1] At lower latitudes, home to most of the Global South's people, rising seas and meteorological volatility will have significant consequences, including food and water insecurity, flooding of coastal cities and towns, and damage to fishing and agriculture—key economic sectors, where productivity may decline by as much as 50 percent in some areas.[2] Further impacts include large population displacements and increased mortality from disease. Inadequate governance, lower technical capacity, and lack of human and financial capital will compound these effects, reducing the adaptive competency of developing nations while increasing their vulnerability.[3]

As defined by the Intergovernmental Panel on Climate Change (IPCC), climate adaptation measures entail "adjustments in human and natural systems in response to actual or expected climate stimuli or their effects that moderate harm or exploit beneficial opportunities."[4] Strategies to adapt infrastructure to the new norms of climate variability vary widely according to asset type, specific risks and vulnerabilities, and local capacity. Moreover, adaptation is rendered more complex by the significant uncertainties surrounding future impacts—the lack of certainty being a "salient characteristic of climate change."[5]

Even in the face of uncertainty, however, attempts have been made to anticipate the economic costs of adaptation. According to a 2010 World Bank study, assuming a temperature increase of 2°C (3.6°F), the annual cost of adaptation in developing nations could be between $75 billion and $100 billion (at 2010 prices). Coincidentally, this is in the same range as the amount of aid that developed countries distribute to developing ones.[6] Another useful metric, cited in the same study, is that the additional costs for climate-proofing new infrastructural assets in the developing world, expressed as a percentage of base infrastructure costs, are in the range of 5 percent to 20 percent.[7]

Much has already been written about policy measures and best practices for adaptation planning and preparedness. The focus here is on climate-compatible strategies that align with the principles of infrastructural ecology. The exemplary approaches covered in this chapter fall into three categories: (1) the use of structural and nonstructural means to protect against riverine flooding and sea-level rise; (2) mitigation of infrastructural damage to critical inland services; and (3) water management in drought-threatened areas—including the containment and reversal of desertification (the drying up of wells and rivers)—and in coastal regions at risk of salt intrusion.

Coastal Protection and Adaptation: Hard and Soft Strategies

Low-lying coastal zones host growing concentrations of populations and economic activities that are at risk from sea-level rise; coastal erosion; and tropical storms of increasing frequency and intensity, along with their accompanying storm surge. Adaptive responses that more flexibly integrate sea and land, incorporating patterns of interaction present in nature, will be increasingly preferable to, and in some cases combined with, constructed measures.

Managing Climate Variability in the Khmer Empire

Adapting large coastal settlements to climate risk is hardly a new undertaking. In order to survive, early empires developed many ingenious solutions to manage the seasonal extremes of both deluges and dry spells. One of the largest settlements of the pre-industrial world was the Khmer empire, which controlled what is now Cambodia from the ninth through the sixteenth century. The vast agrarian domain of Greater Angkor, administered by the Khmer from the capital, at Angkor Wat, supported an estimated one million inhabitants, whose rice-based economy depended on a constructed hydraulic network covering more than 1,000 km² (386 mi²; see figure 5.1). By evening out the

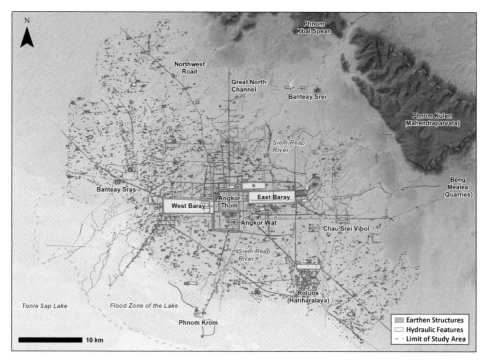

Figure 5.1
Archaeological map of Greater Angkor (Cambodia), circa ninth to fifteenth century. Credit: Damian Evans, Christophe Pottier, and Greater Angkor Project.

effects of oscillating rainfall patterns (six months of flood followed by six months of drought), this network served as a climate-adaptive strategy.[8] At the same time, by redistributing captured and stored surface waters, along with their nutrients, the constructed systems promoted year-round rice production.

The network performed three principal functions: water collection and control, storage, and dispersal and delivery.[9] Stormwater from the northern hills, which would otherwise have overflowed the river delta, streamed into the central region's three large reservoirs (*barays*), for later release into the canal system and agricultural landscape; the water then flowed further southward, where it was used for irrigation and to prevent the intrusion of saltwater into the fields.[10] The *barays* were a major infrastructural intervention; the largest, almost 8 km (5 mi) long and 2.4 km (1.5 mi) wide, were framed by embankments three stories high.[11] Archaeologists believe that, given the seasonal saturation of the fields, the elevated banks of the canals doubled as roadways;[12] one has claimed that they were the equivalent of freeways.[13]

Rice was cultivated in paddies (flooded arable fields),[14] which were supported by an immense system of canals, moats, ditches, dykes, levees, and small dams. During the dryer months, these structures helped to distribute floodwaters laterally. To ensure adequate support for the empire's economy, settlements were widely dispersed, but they remained connected by means of the roads, temples, and bridges laid out atop the raised embankments that rose from the flat terrain of the delta.[15] This dual-purpose water-regulation and transportation system created low-density settlements at such a large scale that it may have contributed to the empire's ecological collapse.[16]

The demise of Greater Angkor, which began in the fourteenth century, has been attributed to a combination of ecological breakdown and infrastructural failure. The first was likely to have been brought about by excessive harvesting of upland forests for fuel and overcultivation of fields; it has also been speculated that weakened monsoonal patterns (characterized by decades-long droughts broken by torrential, destructive storms)[17] and methane emissions from rice cultivation sparked further environmental damage.[18] Among the possible causes of infrastructural failure were the erosion and degradation of topsoil and the accumulation of downstream sediments, both of which may have resulted from the reconfiguration of the river and the development of the canal systems.[19]

The establishment of this immense, integrated network underscores the capacity of preindustrial societies to undertake advanced, multifunctional infrastructural interventions. At the same time, the empire's ultimate demise serves as a warning about the limitations of such prowess in the absence of systems thinking—specifically, the failure to understand the potential ecological implications of deploying such a massive system. This lesson is especially poignant today.

Jakarta's Coastal Defense

Flooding is hardly a new problem for Jakarta, Indonesia, a city of 10.2 million people that is widely considered to be among the most vulnerable to the effects of climate change. Jakarta is situated on a deltaic plain formed by the confluence of 13 rivers, all of which originate in the mountains south of the city. Today, 40 percent of the city's land is below sea level; by 2030, that percentage is expected to double.[20] By 2007, a study had already confirmed a 0.8 m (2.6 ft) seawater rise off Jakarta's coast.[21] Meanwhile, Indonesia's surface temperatures have been increasing between 0.2°C and 0.3°C (0.18 and .36°F) per decade, and, according to one report, "climate departure"—a state in which damaging meteorological events become the new normal—is expected by 2029.[22]

Since its founding as the seat of a fifth-century kingdom, through its colonization by the Dutch (who constructed drainage canals and floodwater storage reservoirs), and down to the present, Jakarta has been at risk from both rain runoff and tidal flooding.[23] Moreover, it is routinely punished by rainy-season inundations that impose high economic and human costs—particularly on the urban poor, who occupy much of the land that is below sea level.[24] More recently, the city's waterways—clogged with refuse and filled with eroded sediment from upstream water sources—have begun to suffer from insufficient drainage and storage capacity, further increasing vulnerability to flooding.

Jakarta also suffers from land subsidence, which is occurring at a rate of between 0.5 and 0.17 cm (0.19 and 0.067 in) per year.[25] Among the causes of this soil consolidation are rapid population growth, which has led to uncontrolled water withdrawals and the intensification of land use (building and road construction). Increased water use, in turn—along with the lack of opportunity for natural water infiltration—has contributed to a shortage of clean water. Finally, the city's loss of green space and its increasing vehicular density have intensified the heat island effect—significantly warmer temperatures caused by dense settlements—compounding health concerns.[26]

Jakarta's complex risk profile calls for multitiered adaptation strategies, both structural and nonstructural. Through a series of coordinated projects, the capital district government, DKI Jakarta, is beginning to integrate climate change mitigation and adaptation into spatial planning while simultaneously attempting to decrease poverty.[27] The Dutch-funded Jakarta Coastal Defense Strategy calls for a replacement seawall to be constructed, at a cost of $2 million, behind a sinking and crumbling one. With the support of donors and the central government, riverbank dike levels are being raised throughout the city.[28] And with financing from the World Bank, rivers and canals will be dredged, increasing their drainage and storage capacity.

To further support storage capacity, two multipurpose tunnels are being planned. Like the SMART system constructed in Kuala Lumpur (see chapter 2), Jakarta's tunnels will decrease congestion by rerouting traffic and, during times of emergency, store monsoonal rains.[29] The Pluit reservoir revitalization project, another effort to increase water storage capacity, will enlarge an existing reservoir and transform its banks into landscaped parks and recreation areas using "blue and green" infrastructure (water-storing plantings). The project will also entail planting 10,000 new trees and relocating 3,000 squatters, formerly housed on the reservoir's banks, to new, subsidized housing.[30] This pilot project will set the standard for future reservoir enhancements.

Perhaps the boldest measure—also a collaboration between the Netherlands and Indonesia—will be a $40 billion, multipurpose land-reclamation project located just off the coast. The reclaimed land, which will take the form of multiple islands, will be integrated with yet another seawall, 32 km (19.9 mi) long, financed by the sale of residential and commercial real estate situated on the islands (the seawall structure will also provide a platform for rapid transit and toll roads). This new enclosure system will create an even more resilient barrier for the city—and, according to the plan, will eventually transform Jakarta Bay into a freshwater reservoir.[31] As of this writing, the project had been put on hold pending further cost and environmental evaluations by the national government. Even without this project, however, the measures currently planned and under way represent the multifaceted solutions that are necessary to secure a future for the nation's economic center.

Reviving Soft Buffers: Mangrove Restoration in Viet Nam

In 1999, a supercyclone that traversed the Bay of Bengal devastated the state of Orissa in eastern India. Winds of over 300 km/h (186 mph) powered a tidal surge of 8.5 m (28 ft), wiping out villages, killing nearly 10,000 people, and leaving behind $5.1 billion in damage. Communities protected by mangrove belts, however, were less affected.[32]

Mangroves provide a host of ecosystem services that, amid the consequences of human-caused climate change, are becoming increasingly valued. As natural buffering systems, mangroves provide coastal reinforcement; that is, climate "regulating services."[33] The combination of sediment deposited from rivers and the mangroves' inherently tough vegetation dissipates wave energy, absorbing as much as 90 percent of it.[34] Mangroves also reduce erosive winds, stabilize the sea floor, and, by trapping sediment, foster land building in a seaward direction. Significantly, mangroves also mitigate climate change by storing carbon.[35]

As intertidal tropical and subtropical coastal wetlands, mangroves also offer a number of "provisioning services" by meeting subsistence requirements for fuelwood, high-quality building timber, thatch, and fodder for domestic animals; and by being sources of nontimber products, such as tannin and medicines. Their flowers provide nectar for honey production, and their cover (leafy shade and thick roots) fosters biodiversity. Finally, as spawning and nursery grounds for fish and crustaceans, mangrove forests support aquaculture-based livelihoods.[36]

Nevertheless, human activities have eliminated as much as 35 percent of the world's mangrove cover.[37] These valuable buffers, which are legally considered commons, have been systematically deforested, both for wood and to support a burgeoning aquaculture economy that supplies a third of global seafood consumption but that has been rendered unsustainable by the introduction of nonnative species, pathogens, and discharged wastes.[38]

Viet Nam has a special history when it comes to mangrove restoration. In the Mekong Delta, after the United States had used Agent Orange, a chemical herbicide, in a systematic defoliation campaign (1965–1969) undertaken during the Vietnam War, the

Figure 5.2
Mangrove plantation, Thua Thien Hue Province, Viet Nam. Photo Credit: Courtesy of the Center for Social Research and Development (CSRD), Viet Nam.

Vietnamese rebuilt some 40,000 ha (98,842 acres) of the delta's mangrove forest.[39] More recently, in northern Viet Nam's Red River delta, the Viet Nam Red Cross (VNRC) has taken up efforts to manage rising sea levels, which have increased by 20 cm (8 in) over the past five decades and are expected to reach 1 m (3.28 ft) by 2100.[40] The delta's crops and settlements were already protected by a sea dike system, but, since 1994, VNRC staff and local communities have planted more than 22,000 ha (54,363 acres) of mangrove forests along a 100 km (62.1 mi) stretch of coast (see figure 5.2). Some 350,000 people are the direct beneficiaries of this protection,[41] and the project was so successful that in 1997 it was expanded to include seven other provinces in northern Viet Nam.[42]

The reestablishment of lost mangroves on the seaward side of sea dikes reduced the cost of maintaining constructed defenses against the five to ten coastal storms that occur every year.[43] It also lessened the damage associated with storms. For example, a level 11 typhoon that struck in 1994 caused $400,000 in damage; and in 2006, an equally powerful storm generated less than half that amount of damage: $180,000. Moreover, private property sustained 90 percent of the damage resulting from the first storm, whereas for the second storm it was 25 percent.[44] The overall value of the risk avoided has been calculated at $15.0 million—almost twice the $8.88 million cost of mangrove planting.[45]

In addition to coastal protection, the restored mangrove forests offer another benefit in that during the last several decades, the restored mangroves in Viet Nam's coastal zone have experienced an upsurge in controlled shrimp farming (farmers can earn five to seven times more per unit of cultivation by farming shrimp than by growing rice).[46] The restoration effort also yielded still other benefits, including disaster-preparedness training for 10,000 individuals and the planting of bamboo and casuarina trees along the banks in riverine communities to reduce erosion.[47] Most important, the carbon dioxide emissions absorbed by the reforested areas are valued at $218 million, assuming a value of $20 per metric ton (1.102 ton) of CO_2.[48]

The VNRC project—as well as another similar initiative, undertaken by CARE—features nature-based adaptations that maintain a careful regard for ecosystem sustainability while extending benefits to local communities.[49] Such efforts reflect a burgeoning global network of citizens and NGOs collaborating on integrated approaches to infrastructural resilience.

Multipurpose Adaptation: The Floating Gardens of Bangladesh

Like Indonesia, Bangladesh is among the nations most at risk from the effects of climate change. Situated on four major river deltas, Bangladesh is subject to rising sea levels and monsoonal flooding. Every three years, on average, a major cyclone creates storm surges more than 10 m (32.8 ft) in height.[50]

Two-thirds of the population of Bangladesh live less than 5 m (16.4 ft) above sea level, and predictions indicate that by 2020, 42 percent of the population—80 million people—could be displaced by rising waters.[51] Waterlogging of its agricultural base, and the resulting reduction in cultivable land, threatens at least one-quarter of Bangladesh's GDP and the livelihoods of 65 percent of its workforce.[52] Between 2005 and 2050, the nation faces an annual anticipated decrease in agricultural GDP of 3.1 percent.[53] The southern and northwestern regions are the most vulnerable to crop loss and associated food insecurity. In the south, saline intrusion into coastal groundwater further threatens the agriculture of the rural poor. In this fiscally constrained country, the development of climate-resilient crops, livestock, fishery systems, and disaster sheltering is imperative to offset the future effects of climate change.[54]

In an indigenous technique that has been practiced for hundreds of years in the flood-prone and waterlogged districts of southwestern Bangladesh, rice straw is used to construct production beds (*baira*)—long, rectangular gardens that float in rivers, canals, lagoons, and wetlands. These flexible (rising and falling) fields are capable of riding out any inundation and can be accessed by farmers in small boats. Made of virtually free materials and fabricated using local knowledge, the *baira* may be the forerunners of hydroponic farming.

Figure 5.3
Floating vegetable gardens, village of Mugarjor, Pirojpur District, Bangladesh. Photo Credit: Mustafiz Mamun.

Among the communities that have adopted *baira* as an adaptive strategy, more and more are selecting water hyacinth, an invasive weed, as the matrix of choice for the floating platforms. Farmers construct mats from mature water hyacinths, which are then overlaid by bamboo poles. To form compost, the farmers then lace decaying vegetation, such as duckweed, into the mats, often adding cow dung. This base then receives vegetable seedlings. Typically, vertical bamboo poles are used to fix the *baira* in place. *Baira* can be cultivated in waterlogged, otherwise fallow areas, producing two to three harvests of mixed crops year-round (much more than land-based vegetation), and without need of chemical fertilizers or irrigation (figure 5.3).[55] *Baira* are used to produce okra, cucumbers, eggplants, beans, tomatoes, cabbages, red amaranth, pumpkins, and potatoes, among other vegetables, as well as a multitude of spices.[56]

Bangladesh's floating agricultural system serves many purposes. During floods, *baira* have been used as refuges for poultry and even cattle. The platforms attract fish, which supplement farmers' incomes. After cultivation, the excess biomass can be used as organic fertilizer in nearby land-based fields. In the northwestern part of the country, under the aegis of several NGOs—including CARE and the International Institute for Environmental Development (IIED)—this highly flexible technology is being further developed as a community-based adaptation, meaning one that is capable of being instigated and managed locally.[57]

In China, floating platforms are being considered not only for agricultural production but also for environmental protection. Some 57 percent of China's 40 principal freshwater lakes are experiencing eutrophication (deterioration into algae-laden bogs) from agricultural runoff or untreated sewage, an ecological setback that will only increase with rising temperatures. To address this problem, scientists are reintroducing artificial floating islands (AFIs)—buoyant frames, constructed of various materials, which are similar to those that were used 1,700 years ago for rice and vegetable cultivation. The microorganisms that colonize the islands purify the wastewater by introducing dissolved oxygen and reducing nutrients.[58] When coupled with other technologies modeled on natural systems, AFIs have the capacity to treat sewage to meet current standards for nonpotable uses. At the same time, the islands can provide spawning beds for fish and can stabilize and protect littoral zones from the impact of waves.[59]

Despite the romance of these versatile adaptive measures, experts are cautious about the viability of using floating gardens in areas near coastlines. Water hyacinths cannot flourish in brackish water or in waterlogged areas subject to salt intrusion, nor can they survive where rainfall is insufficient. To determine how this technology will fare in the long term, under climate change, and what further adaptations may be needed to compensate for its shortcomings, more research will be needed.[60]

A Floating City: Makoko, Nigeria

Sustaining settlements and livelihoods in areas that are susceptible to rising water levels will demand a range of adaptation strategies, including those that incorporate flexible relationships between water levels and critical services. One of the world's most ambitious water-bound settlements is the Netherlands' Waterbuurt, a community of one thousand residents that is entirely afloat on the waters of Lake IJ, in the East Amsterdam neighborhood. Moored to submerged concrete tubs, the homes of Waterbuurt rise and fall with the tides.

Coastal communities everywhere face two choices: move residents upland, away from valuable yet vulnerable properties, or live with water, adapting by means of floating or amphibious structures. In the Global South, a number of innovative responses to this dilemma are emerging.[61]

Africa's coastal towns—among the most urbanized on the continent—face imminent risk of flooding from a combination of rising sea levels and climate variability. In Lagos, Nigeria, the world's fifth-largest megacity, slum dwellers—most of whom immigrated to the city from rural areas over the past several decades—make up 70 percent of the population. Many of the city's poorest residents live in some of the oldest settled areas: the hazardous, marshy regions along the coast and the bay. Makoko (population 100,000), one of nine squatter communities in this area, is sited on Lagos Lagoon, where an eighteenth-century fishing village once stood. The settlement is conveniently close to downtown, but residents live mostly in one-room structures built on stilts on reclaimed land—sites highly susceptible to tidal and rainy-season flooding.[62]

In Makoko, where it is possible to earn a living through mariculture or by working in nearby sawmills, newcomers have found the answer to the city's overpriced real estate. But unregulated expansion, lack of sanitation and water infrastructure, polluted runoff from impervious surfaces, and a problematic land-titling process have created health concerns and insecurity for the Makoko community while dissuading private investment.[63] Because of Makoko's prime location, the Lagos State government created redevelopment plans for the settlement that called for upscale entertainment, recreation, and tourism. As recently as 2012, Makoko's residents faced eviction—without compensation.

In the meantime, however, in a project conceived by NLÉ architects, in collaboration with the Makoko Waterfront Community, local residents rebuilt a compromised schoolhouse as a lightweight timber structure that floats on recycled plastic barrels, its bottom-heavy design ensuring stability (see figure 5.4).[64] Developed through a participatory design process and relying on local construction techniques, the three-story structure has ventilated bamboo walls and a sheet-metal roof; because it is oriented east to west to capture prevailing winds and reduce solar gain, it is perfectly adapted to its microclimate.[65]

In addition to its advantageous design features, the building offers internal infrastructure services. It is powered by solar photovoltaic panels (565 W) and a battery bank. The large roof-catchment area is used to harvest rainwater, which is supplemented by solar distillation of lagoon water. Composting toilets provide safer sanitation than land-based facilities, as Lagos lacks a central sewage system. Part boat, part building, and part infrastructural facility, the structure is adapted not only to rising water levels but also to changing community needs.

The threatened removal of this highly publicized project heightened global attention to Makoko's plight. It also gave rise to the Makoko/Iwaya Waterfront Regeneration Plan, which calls for the community's vulnerabilities to be transformed into strengths.

Figure 5.4
Makoko floating school, Lagos, Nigeria. Photo Credit: Iwan Baan.

In October 2013, with support from the Social and Economic Rights Action Center (a Lagos-based NGO), as well as numerous additional stakeholders, residents submitted the plan to the Lagos Ministry of Physical Planning. Built on a self-help model designed to foster community renewal,[66] the plan features proposals upgrading low-income housing, providing infrastructural and health services, and promoting tourism and commercial development. Public services, for example, would take the form of shared neighborhood "hotspots"—flexible, decentralized facilities providing energy, potable water, sanitation (toilets and showers), and waste collection. The hotspots would be powered by a resident-operated biogas cooperative fueled by organic and human waste, with the by-product to be used as fertilizer for agriculture.

The plan presented a powerful enough vision that in April 2015, in an about-face decision, the Planning Ministry announced its own regeneration plan, an alternative development called Houses on Water, which would incorporate aspects of Makoko's widely publicized floating building. The development would also include schools, a hospital, commercial entities, and fish-processing plants.[67] According to NLÉ's principal, Kunlé Adeyemi, the decision represented an "open-minded reconsideration of established policies."[68] It remains to be seen, however, whether inventive, practical, and self-sufficient models such as Makoko's will become the new norm in flood-threatened tropical and subtropical coastal areas.

Addressing Urban Flooding in Kibera, Kenya

Kibera, one of the world's largest informal settlements, is located in Nairobi, Kenya, along the Ng'ong River, just upstream from the Nairobi Dam. This low-lying area was originally a forest area set aside to protect the watershed of the Nairobi Dam but eventually became a desirable site for informal development because of its close proximity to downtown Nairobi (see figure 5.5). Over the years, two factors—rapid urbanization throughout Nairobi and increased rainfall intensity during East Africa's rainy seasons[69]—have intensified flooding in Kibera, impacting some of the poorest people in the world.

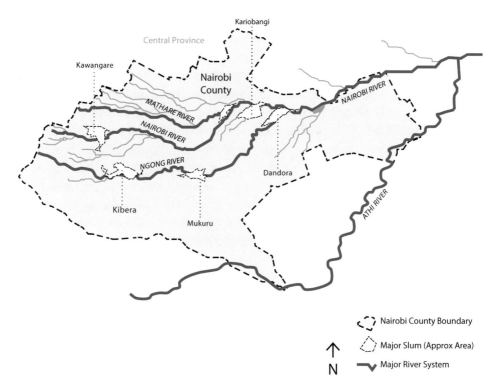

Figure 5.5
Slum areas of Nairobi, showing Kibera's river locale. Credit: Jessica Bremner and Joe Mulligan, Kounkuey Design Initiative.

In May 2015, record rainfall pummeled the city, bringing all of Nairobi to a stand-still, crippling roads, bridges, and other critical infrastructure. Thousands of properties were damaged, and at least 11 people lost their lives.[70] In Kibera, sewage overflowed into the floodwaters and spread out across the community. According to a postflood survey conducted by Kounkuey Design Initiative (KDI), a nonprofit based in the United States and Kenya that has been creating productive public spaces in Kibera for over ten years, 60 percent of nearly 1,000 households surveyed reported outbreaks of cholera in their community.

Working closely with communities along the riverbanks, KDI led an integrated flood-mitigation process in Kibera. KDI often includes some sort of flood protection along the river's edge in its projects, but after the disastrous flood of 2015, it set the additional goals of increasing broad community awareness of flooding risk and demon-strating the various design approaches that can be used to reduce such risk.[71]

The multisector interventions entailed improvements to the river's edge, upland drainage, wastewater collection, and solid waste collection.[72] To increase the size of the river's flow area, KDI constructed a channel alignment with gabions, which creates distinct river flow paths for low, rainy-season, and flood-condition flows. As part of a broader, watershed-based approach to flood reduction, water retention and rainwater har-vesting strategies were implemented in upland parts of the community. A new sewer trunk line was built along the river by Nairobi City Water and Sewerage Company, which provided sanitary drain connections to nearby homes, thereby reducing the sewage flows into the river. In addition, the government implemented a new trash-disposal policy that incorporated household bins, community bins, and a collection routine to keep solid waste out of the river and prevent it from obstructing its flow. KDI's approach was not simply to assist Kibera in rebuilding after disaster but also to mitigate the combined effects of urbanization and climate change, creating productive public spaces that will reduce future flood risk while addressing broader community needs and lessen the gap between the formal and informal areas. Figure 5.6 illustrates the resulting combined activities of community involvement in making public space and protecting against floods, thereby creating successful, active spaces in informal developments.

Inland Adaptations

Perhaps because of the noticeable physical changes and impacts on settlements and infra-structure, coastal cities and states worldwide became attuned to the perils of climate change before inland areas did. Though perhaps less dramatic, the risks associated with

Figure 5.6
Productive Public Space 007 in the informal settlement of Kibera in Nairobi, Kenya. Photo Credit:
Pascal Kipkemboi, Kounkuey Design Initiative.

changing inland temperatures and precipitation patterns are equally serious. Particularly
in areas subject to the urban heat island effect, severe downpours and pavement-melting
heat waves pose equal threats to infrastructure. Some forward-thinking governments
are responding accordingly.

Delhi's Urban Heat Island Policies

In 2015, India was struck by a heat wave that ultimately claimed over 2,500 lives.[73]
Although not all of the deaths occurred in cities, urban areas were hit especially hard
because of the heat island effect. With global temperatures on the rise, urban residents
in both arid and humid subtropical regions are at risk from the compound effects of
climate change and heat islands.

In the world's megacities, effective mitigation of rising temperatures calls for a multisectoral approach to reducing solar absorption, one that integrates transportation planning with urban planning, including the design of buildings and streetscapes. Through
strict enforcement of its 2007 Energy Conservation Building Code (ECBC), New Delhi,

India, is taking just such a holistic approach. Under the ECBC, new buildings must be designed with cool roofing technologies, and building plans that lack provisions for cool roofs will be rejected.

Whereas traditional black roofs can reach temperatures as high as 77°C (170°F) and reflect only 5 percent of sunlight, "white" or "cool" roofs can reflect up to 80 percent of incoming sunlight, resulting in a considerably lower temperature of 43°C (111°F).[74] Cool roofs can be created through the addition of a thick layer of white paint or the application of white roof tiles; solar-reflective coatings and lime coatings are other options. Regardless of the approach, under city law, flat roof surfaces must have an initial solar reflectance of no less than 70 percent and an initial emittance (energy radiated by a body's surface per second, per unit area) of no less than 0.75.[75]

By reducing buildings' overall energy demand, the ECBC also addresses another major contributor to the heat island effect: the waste heat generated by air-conditioning units. ECBC-compliant buildings use 40 percent to 60 percent less energy than their conventionally built counterparts, and much of that reduction comes from more efficient air conditioning (and lighting) systems. Thus, they generate less waste heat from air conditioning. (A 30 percent to 40 percent reduction in air-conditioning demand, for example, yields a corresponding reduction in waste heat.)

Delhi is also addressing another contributor to escalating temperatures: lack of vegetation. Because plants use evapotranspiration to limit the discrepancy between their surface temperature and that of the surrounding air, vegetation generally maintains a lower temperature than building materials. Under the Delhi Street Design Guidelines adopted in 2009, trees must be planted along all major streets in the city[76] and must be located at the street edge to maximize canopy coverage of both the roadway and the sidewalk.[77]

The Delhi Street Design Guidelines also address urban geometry—the height and spacing of buildings. The relative dimensions and positioning of buildings influence wind flow, energy absorption, and the ability of various surfaces to project radiation back into the atmosphere.[78] When buildings are obstructed by larger neighboring buildings, they cannot readily reflect heat back into the atmosphere, especially at night. By fostering better wind flow, buildings constructed at a variety of heights help mitigate this "urban canyon" effect. Orienting buildings to capitalize on the direction of local wind flow also helps keep buildings cool and reduces the need for air conditioning.

To address rising temperatures from a more operational perspective, Delhi is looking across the subcontinent, to Ahmedabad, which in 2013 implemented the Ahmedabad Heat Action Plan. The plan's primary objective is to alert those who are most at risk of heat-related illness when extreme heat conditions are imminent and to help them find protection from the heat.

In developing nations, the combination of rapid urbanization and a warming climate is exacerbating the heat island effect. According to current predictions, global temperatures will increase from 0.14°C to 1.1°C (0.25°F to 2°F) per decade,[79] amounting to an increase of 2.0°C to 3.3°C (3.5°F to 6°F) over the next century. In cities, however, where warming temperatures will be exacerbated by the heat island effect, temperatures could rise between 4°C and 7°C (8°F and 12°F) in the next century. Heat islands are worrisome because of their capacity to intensify climate-induced risks to human health as well as to transportation, water supply, and other infrastructure systems. The harms are also disproportionate in that although heat islands have the potential to affect all city residents, some segments of the population—including low-income, homeless, and elderly residents—are especially vulnerable. Fortunately, cities like Delhi are demonstrating that it is possible to tackle the challenges of climate change in a holistic way.

Rural Road Resilience in Cambodia

Nearly 80 percent of Cambodia's land area falls within the Mekong River and Tonle Sap basin, rendering the nation vulnerable both to annual cycles of drought and flooding and to the effects of climate change. The depth of Cambodia's vulnerability was highlighted in 2009, when Typhoon Ketzana caused an estimated $132 million in damage.[80]

Although the nation's agricultural ecosystems rely on the abundant monsoonal rains, they are ill equipped to handle the extremes—heavier rainy seasons and more severe dry-season droughts—predicted for the future. Moreover, nearly 80 percent of Cambodia's poor live in rural areas,[81] where only 11 percent of roads are paved. And despite the rural economy's dependence on the national road network, the condition of rural roads is only worsening—partly because of a lack of funding for proper maintenance and partly as a consequence of increased traffic brought on by economic development. Thus, Cambodian farmers have been subject to the dual vulnerabilities of low crop yields caused by lack of rain and of difficulty selling their crops when unpaved roads are washed away.

In 2014, the Asian Development Bank (ADB) spearheaded the Rural Roads Improvement Program, which had three related goals:

(i) ensuring transport mobility and access to critical services during extreme events, particularly floods;
(ii) ensuring safe access to water during climate extremes, such as drought; and
(iii) ensuring sustainable and diverse rural livelihoods to reduce vulnerability to climate change.[82]

One of the project's most effective efforts entailed planting trees and grasses that are resilient to climate change along all roadside embankments during road construction. The plantings will increase soil stability and ground cover under and around new roads, enabling them to last longer and making them less likely to wash out during heavy rains. The added vegetation also reduces erosion and improves soil structure for nearby agriculture. Today, both men and women can be found working along Cambodia's rural road network—paving thoroughfares, building stable road foundations, creating drainage pathways, planting trees, and sowing seeds.

Other, more ambitious strategies are designed to store water for the several months that elapse between the wet and dry seasons. These approaches were based on preproject surveys showing that approximately 52 percent of respondents ran out of water during the dry season and that 30 percent could have used more water for livestock, fishponds, and small-scale irrigation. New infrastructure to conserve and store water during the rainy season relies on an integrated network of wells, ponds, and canals[83]— systems that hark back to those of ancient Angkor Wat. Postproject surveys showed that approximately 80 percent of respondents saw the benefit of planting grasses, shrubs, and trees to restore soil stability and to provide shade and fruits.

The program's unique, infrastructural ecology approach also affords social and economic benefits, including improved vehicle access to markets and services for the rural poor, improved access for emergency vehicles during disasters, and support for income diversification in times of climate disturbance.[84] Moreover, such road construction programs generate jobs for low-skill laborers as well as revenue opportunities for small local businesses.

Cross-Sector Solutions for Water Security

As temperatures in developing countries increase, the availability of water from river flows, rainfall, snowmelt, and ground sources may decline, affecting entire ecosystems and their human inhabitants.[85] And as sea levels rise, the aquifers on which coastal cities and megacities depend can become contaminated by saltwater intrusion. Both trends will have disproportionate effects on the poor and vulnerable.

Given the extent to which multiple human endeavors—from irrigation, to industry, to energy generation—depend on the hydrologic cycle, it is imperative to integrate climate adaptation into new development. Although nonstructural approaches to water management, such as conservation and demand modification, are key means of addressing water shortages in the developing world, the focus here will be on sustainable water management and new water sources.

The nexus of energy, water, and food in the context of climate change has become an increasing focus of the development canon.[86] As climate variability yields more frequent and more intense flooding and drought, interactions between the elements of this nexus are undermining the already tenuous reliability of food and energy production and exacerbating competition among all sectors for access to water. For example, water stored and released for electricity generation can yield drought or deluge, impacting irrigation cycles, and the energy demands generated by wastewater treatment, water distribution, and irrigation pumping can tax power plants—which, in turn, rely on water for cooling.

Infrastructural solutions in emerging economies must reflect the complex interdependencies of food, energy, water, and climate—and take advantage, where possible, of advantageous synergies. The goal is to achieve multisector benefits while averting unintended consequences.

Sustainable Water Management: Reviving Ancient *Amunas* in Lima

Lima, Peru, is the world's second-largest desert city. As the highlands of the Andes succumb to greater dryness from climate change—brought about by a combination of glacial retreat and reduced snowpack—Lima finds itself once again tapping the watersheds above the city to serve its approximately nine million inhabitants. In the surrounding mountains, pre-Incan aqueducts will be restored to meet the city's burgeoning needs.

Typically, during the four- to five-month wet season, the city sustains a water surplus: its rivers flood, and landslides are common.[87] During the dry season, however, Lima is short on water and must draw down the main aquifer beneath the city (which is fed by the Rimac, Chillon, and Lurin Rivers) as well as constructed reservoirs. Despite transporting additional waters from the headwaters of the Amazon basin, the city still experiences an annual deficit of approximately 43 million m^3 (11.35 billion gal).

The area's early settlers, the Wari, were small-scale farmers who flourished from AD 500 to 1,000. Like other Andean cultures, the Wari had to deal with seasonal variations in water availability. To do so, they devised a complex water-conservation system that relied on terraces sculpted into the steep terrain to capture rainy-season water that would otherwise have run swiftly into the sea. Stone canals (*amunas*) transferred the water laterally across the slopes to areas where it would slowly percolate through the soil into groundwater. After several months, the stored water would be slowly released to surface downstream, where it would feed irrigation channels.[88]

Over time, the Wari became masters of water control, even withstanding a 30-year drought that occurred around AD 600.[89] After the Wari state disappeared, its system of terraces and *amunas* provided valuable support for the subsequent Incan civilization, which expanded the Wari's hydrotechnology across the rest of the Andes (see figure 5.7).[90]

In the spring of 2015, Lima's government announced that it would begin using watershed management to regulate both surplus and scarcity. As a first step, hydrologists and project developers from SEDEPAL, Lima's water utility, toured the uplands, in the company of Andean community leaders, who showed them the ancient *amunas*, long fallen into disrepair. The interventions outlined by Forest Trends, a Washington, D.C.–based NGO that is partnering on the project, will include repairing and regrouting the *amunas* so that the ancient systems can once again convey water to various locations for seasonal storage. Although some water will be diverted for local irrigation, the rest will be used to recharge local groundwater, amplifying the aquifer's base flow.[91]

This revival of ancient hydrologic technologies will be accompanied by efforts to increase the catchment and storage of upland rains—principally through the regeneration of the Puna grasslands. In these areas, which are located in the high altitudes of the rivers' watersheds, the loss of grass cover has led to erosion, and cattle have compacted the soil. To regenerate the water-retaining grassland ecosystems, key areas will be protected from grazing, and rotational grazing will be used in others.[92]

The hydrologic feasibility report estimates that the canal refurbishment and green infrastructure strategies could yield an annual increase of roughly 40 million m^3 (10.56 billion gal), virtually erasing the dry-season deficit. At an estimated cost of $23 million, the report suggests that these measures—totaling $0.51 per m^3 (2 cents per gal) of water—will be far more cost-effective than building a desalination plant, which would provide equivalent flow at a cost of $0.73 per m^3 (2.8 cents per gal).[93] As in other parts of the world—including India's Godavari basin, discussed in chapter 3—time-honored indigenous water-management measures may outperform industrial-era technologies.

Water Banking in Kazakhstan's Fergana Valley

Many developing regions experience sharp competition for water to support energy uses, irrigation, and domestic and industrial uses. If water is shared equitably both upstream and downstream, multiuse dams may serve both needs; however, many rivers cross national boundaries, adding complexity to efforts to ensure equitable use.

In Central Asia's Fergana Valley, the Syr Darya River and its tributaries stretch across Kazakhstan, Kyrgyzstan, Tajikistan, and Uzbekistan. Fed by mountain glaciers and snow that will ultimately be diminished by climate change, the Syr Darya River is currently used for both hydropower and irrigation, thanks to a series of multipurpose dams. Until the 1960s, the river's natural flow sufficed for irrigated agriculture, with high flows in

Figure 5.7
Restoration of Incan *amunas*, Tipón, Peru. Photo Credit: McKay Savage, Wikimedia Commons.

summer and low flows in winter. But in the decades that followed, the amount of land under cultivation effectively doubled, from 456,000 to 977,000 ha (1,127 to 2,414 acres)—increasing irrigation demands and contributing to the infamous shrinking of the Aral Sea.[94]

In the early 1990s, to optimize hydropower production, the main releases of water from Kyrgyzstan's reservoirs were shifted from the summer to the winter months, a change that increased hydropower production by 30 percent but created a 20 percent annual shortage in irrigation water downstream (in Kazakhstan and Uzbekistan). This, in turn, led to a 46 percent decrease in cotton yields because of shortages of irrigation water and increasing salinity.[95] Meanwhile, the water from the extra winter releases was diverted into another basin.[96]

The Fergana Valley has long been the breadbasket of a vast transboundary region covering 22,000 km^2 (8,500 mi^2). Recognizing both the vulnerability of surface sources, in the face of climate change, and the complexities associated with transboundary water sources, water experts have been recommending irrigation practices that instead of relying solely on surface water also draw on groundwater aquifers. In the regions of the Fergana Valley that lie within Uzbekistan, the International Water Management Institute (IWMI) has been conducting pilot experiments in the Fergana Valley, in which some of the winter releases are being diverted to local aquifers, where they are "banked" for use during the summer months, when demand is at its peak.[97] Water banking allows water resource allocation to better align with the principles of equity, efficiency, and sustainability.[98] Because groundwater is more reliable under conditions of climate variability, shifting toward combined use of river water and groundwater is a valuable adaptive strategy.

Reclaiming Wastewater for Agriculture in the Jordan Valley

Wastewater, typically regarded as a problem, can also be a valuable resource. As demonstrated by the Chinese dyke fishponds (chapter 2), and the *chinampas* of Mexico and the East Kolkata Wetlands (both in chapter 3), the practice of applying wastewater to land has been and remains widespread, particularly in developing countries. However, of all the land worldwide that is irrigated with wastewater, only about 10 percent is irrigated with *treated* wastewater.[99]

In addition to its most prevalent use, in agriculture, wastewater can be reused in landscaping and aquaculture, to improve wildlife habitat, and to recharge groundwater and aquifers.[100] Irrigation with wastewater (often combined with greywater) closes the resource loop by delivering valuable nutrients and has long been associated with improved yields and livelihoods.[101]

Today, treated wastewater sludge is increasingly being applied worldwide, in both low- and middle-income countries—partly because of high disposal and landfill costs but primarily because of its ability to enhance soil quality.[102] Nonetheless, the safe recycling of wastewater for agricultural use continues to be controversial in fast-growing economies, where successful application of the approach demands tightly controlled conditions capable of preventing passage of disease-causing microorganisms.

As climate change bears down on the already water-scarce Middle East, projections indicate that by 2025, Jordan's annual renewable water resources per capita will fall from a low of 160 m^3 (42,267.5 gal), recorded in 2000, to 90 m^3 (23,775.5 gal)—far below the 500 m^3 (132,086 gallons) defined as "absolute scarcity."[103] The drop stems from a combination of factors: a surging population (including an influx of refugees), urbanization, higher living standards, and climate perturbations. With irrigated agriculture in the Jordan Valley consuming as much as 42 percent of available freshwater sources,[104] the nation had to look elsewhere. Among alternative water sources, wastewater proved to have the lowest marginal cost.[105]

Since 2003, through the Reclaimed Water Project (RWP), the Jordan Valley Authority (JVA) has set its sights on the large-scale reuse of reclaimed water for agriculture. An international effort budgeted at $13.5 million, the RWP is under the general oversight of the JVA and is being implemented with technical expertise from Deutsche Gesellschaft für Technische Zusammenarbeit GmbH (GTZ) and funding from the German Federal Ministry for Economic Collaboration and Development. With the goal of recovering approximately 66 percent of the 250 million l (66 million gal) of wastewater released throughout the nation each day, the JVA has adapted and upgraded its 22 wastewater treatment plants, most of which merely provided waste-stabilization ponds (holding basins where organic matter decomposes biologically).[106]

Since the upgrades, which are in accordance with sound environmental and public health regulations and practices, wastewater is now treated primarily by "activated" (mechanically aerated with oxygen) sludge systems. The wastewater is then fed into a main canal—through which it travels the 100 km (62.1 mi) length of the Jordan Valley, driven by gravity. Along the way, it is mixed with greywater or freshwater and delivered to approximately 4,000 farms over a 10,000 ha (38.6 mi^2) area through a drip-irrigation process that reduces evaporation. (To further limit evaporation and control pathogen contamination, the crop rows are covered with thin plastic sheeting.)[107]

In 2006, GTZ and the JVA promulgated quality guidelines for irrigation water, along with measures for monitoring crop quality. Since then, intensive training has been used to help disseminate and apply the guidelines. In 2009, the guidelines were issued in the form of an advanced digital information system that helps farmers optimize "fertigation" (irrigation and fertilization) according to water quality, location, soil type, and crop and consumption type (fodder, fruit, vegetables intended for cooking, and vegetables intended to be eaten fresh).[108] Because crops are exported to Europe and must meet stringent standards, this system is critical.

The reuse of wastewater for irrigation offers significant benefits. First, as of 2009, when compared with the cost of imported fertilizer, nutrients obtained through wastewater saved Jordanian farmers about JD 4.0 million ($5.6 million) annually.[109] Second, chemical fertilizers are associated with the risk of soil salinization. Third, manufactured fertilizers are estimated to consume about 1.2 percent of world energy—and are thus responsible for an equivalent proportion of greenhouse gas emissions.[110]

Nevertheless, the wastewater reclamation project is not free of problems.[111] Although the wastewater supply had historically included little industrial discharge, that pattern is beginning to change as Jordan attracts more industry.[112] This poses the risk that chemical pollutants could enter the food chain.

Restoring Land and Water through Planned Grazing in Zimbabwe

As of 2000, 70 percent of worldwide agricultural areas consisted of 3.5 billion ha (8.64 billion acres) of arid and semiarid rangelands—essential grasslands that make up one-quarter of the world's land area, harbor approximately 20 percent of the global carbon stocks held in soil, and support roughly 20 percent of the world's population.[113]

In these landscapes, where rain is both spatially and temporally variable, water is a limiting factor. Scientists are now learning that in grasslands, environmental degradation (both from climate change and human activities) can short-circuit the water cycle, leading to desertification.[114] On the human side, degradation stems from poor grazing and cultivation practices as more and more people attempt to gain their livelihoods from livestock management.[115] In a positive feedback loop—one that amplifies changes already set in motion by poor grazing practices—grasses fail to recover, and degraded soils retain less water. This, in turn, leads to the loss of water-holding organic nutrients, which become windblown sediment. Hoping for their pasturelands' recovery, many land managers have reduced the size of their herds, increased the rest cycle in grazing lands, or used fire. But these practices have fallen short of addressing their needs.

Enter Allan Savory, head of the Savory Institute in Zimbabwe and Albuquerque, New Mexico, who is singlehandedly changing our understanding of global grasslands. Savory's observation of ecological dynamics gave rise to the insight that wildlife and grazing animals actually serve to maintain the grassland landscape. As a biologist and holistic thinker deeply immersed in local ecology, he understood that many deep-rooted grass species had coevolved with herbivores. Clustered together against predators, large herds would migrate across the rangeland—consuming the grass, breaking up and aerating the soil with their hooves, and fertilizing the soil with their dung and urine.[116] Savory believed that the timing and cycling of such "disturbances" was key to soil health, water infiltration, and the recovery of the perennial grasses.

The Africa Centre for Holistic Management (ACHM), which was established by a group of Zimbabwean tribal chiefs, hopes to restore livelihoods on 8,000 ha (19,768 acres) of state and private land known as the Dimbangombe, which includes a private parcel donated in 1994 by Savory; the Dimbangombe makes up a small part of the 405,000 ha (over a million acres) of Hwange communal lands. In 2005, the Dimbangombe College of Wildlife, Agriculture and Conservation Management was accredited to provide local communities with vocational land-management training and was granted its own 5,260 ha (13,000 acres) of land.[117]

In a 2008 pilot project on the Dimbangombe, Savory recreated the "critical symbiotic relationship between predators, herbivores, plants, micro-organisms, soils, water and weather" by increasing the herd size from 100 to 600, penning the animals, and rotating their location to new ground weekly.[118] By the following year, grass forage yields had quadrupled and the remaining plant debris cover had doubled, reducing bare ground by 31 percent and soil loss by 17 percent.[119] Most important, the Dimbangombe River began to run again during the dry season; where it pooled, creating new open water, fish and water lilies returned (see figure 5.8). By 2010, the planned grazing techniques had reduced bare soil by 52 percent.[120]

Savory had long been shunned by the scientific community for his counterintuitive speculations and "heretical" practices, but his results have held up and have been widely replicated. To date, Savory's approach is being used to manage only 8,093 ha (20,000 acres) of Hwange's communal lands but, thanks to the Savory Institute, which promotes grassland restoration, the practices have spread to other parts of Africa as well as to Australia, Canada, Mexico, and the United States, and 12 million ha (29.6 million acres) of rangeland are currently under holistic grazing practices.[121]

Using very low-input technology, Savory has demonstrated that manipulating the impact of animal grazing can restore ecosystem health. Savory's research has shown how insects, worms, and fungi transport nutrients and synthesize chemicals that store water and significantly contribute to the creation of carbon molecules that build up in the soil.[122] According to Savory, holistic planned grazing of grasslands can be a key factor in carbon mitigation. Acknowledging that corroboration is required, he has roughly calculated that the world's 5 billion ha (12.35 billion acres) of degraded grassland could remove 12 metric gigatons (13.2 gigatons) of atmospheric carbon annually, returning it to the soil sink. This is equivalent to 6 parts per million (ppm) in carbon

Figure 5.8
Restoration of the Dimbangombe River, Zimbabwe. Photo Credit: Courtesy of Pamela Ngwenya.

reduction each year, which, after the subtraction of current emissions (2.5 ppm/year), would yield a net annual improvement of 3.5 ppm. Savory believes that a concerted effort in this direction could return the planet to preindustrial CO_2 levels in less than 40 years.[123] Even discounting the potential magnitude of carbon storage—which has few, if any, downsides—planned grazing is a simple and promising activity with many benefits, including livelihood restoration, hydrologic improvements, and ecosystem regeneration.

Hybrid Power Plants for Renewable Energy and Desalination
Many parts of the Middle East and North Africa (MENA) are experiencing rapid population growth and improvements in living standards; in these locales, energy demand is increasing, and freshwater supplies are being drawn down at an alarming rate.[124] Detailed climate analyses for the MENA suggest that the combined effects of evapotranspiration and changes in rainfall patterns will reduce renewable water resources by 12 percent each year.[125] Conservation, reuse, and desalination will be critical to maintaining adequate supply.

By lowering the percentage of dissolved salt and minerals in water, desalination can render brackish surface water and groundwater, as well as seawater, usable for other purposes. Although technologies vary, the two basic approaches to desalination are thermal evaporation and reverse osmosis;[126] both separate seawater into freshwater and concentrated brine.[127]

As of 2013, more than 17,000 desalination plants were scattered across 150 countries, with a capacity totaling more than 80 million m^3 (21.13 billion gal) per day.[128] Since close to 40 percent of the world's population live less than 100 km (62.1 mi) from the sea, this source—being independent of climate change—could be considered sustainable.[129] However, desalination has two disadvantages: first, it is notably energy intense; and second, because desalination plants regularly discharge concentrated brine (as well as other contaminants, from facility corrosion and cleaning), they place a heavy burden on coastal waters, sediments, and marine organisms.[130]

Although it constitutes less than 1 percent of global desalination capacity, small-scale, direct solar desalination can be undertaken in areas with high insolation potential by means of solar stills that evaporate and then condense seawater. Indirect solar desalination relies on photovoltaic or fluid-based thermal collectors, which provide heat or power to a conventional desalination plant.[131]

In 2017, the world's first utility-scale solar photovoltaic desalination plant is expected to be completed in Saudi Arabia, where it will supply 60,000 m^3 (15.8 million gal) of water annually to the city of Al Khafji. The $130 million plant is being constructed by the commercial arm of the King Abdulaziz City of Science and Technology, which has commissioned Abengoa, a Spanish renewable energy group, to build a 15 MW dedicated solar-tracking plant to power the desalination.[132] The project is a realization of Saudi Arabia's intent to largely replace its oil- and gas-fired generators with solar and nuclear energy. Although the project underscores the feasibility of renewably powered desalination, its capital-intensive technology would have no purchase in most emerging economies.

Abengoa, however, is pursuing more affordable solutions elsewhere. According to experts, Chile—with its windswept coastline and vast interior deserts—has the largest potential renewable energy resources in the world.[133] In fact, Chile was the home of the earliest solar-powered desalination plant, which was constructed in 1892.[134] The small Lluta Valley plant—located in the Atacama Desert near Arica— cost $210,000 to construct and uses a low-energy membrane-separation technology (brackish water forced under high pressure) powered by photovoltaic panels. In addition to salt, the process removes boron (an element that can be toxic even at low concentrations) from the desert water. Constructed by Fundación Chile, a nonprofit technological-innovation corporation, the plant is located in a parched region where farmers struggle with insufficient water for irrigation; replication and scale-up are planned in the vicinity.

The Seawater Greenhouse: Coproduction of Food and Water in the Desert

If, according to global averages, a person's daily drinking water needs are between 2 to 4 liters (.5 to 1.05 gal), the water required to produce their daily food requirement is between 2,000 to 5,000 liters (528 to 1,320 gal), roughly a thousand times greater.[135] Thus, even when powered by renewable energy sources, desalination can be no more than a partial response to a water-short planet. Throughout the world, but particularly in desert regions, a more secure future will require radical reductions in use of both energy and water. And achieving these reductions will depend on the widespread use of closed-loop solutions.

As it turns out, saltwater, carbon dioxide, and desert sunlight are the primary inputs for the Seawater Greenhouse (SWG)—a bold, low-impact invention that capitalizes on the interdependencies of food, water, and energy. Within a cool, moist, enclosed environment engineered for intensive crop cultivation, the SWG transforms seawater into freshwater for irrigation (see figure 5.9). According to its inventor, British entrepreneur Charlie Paton, the freshwater surplus produced through the distillation cycle is five times the amount of freshwater required by the greenhouse crops and is thus available for other uses.[136]

In this ingenious, solar-powered arrangement, desert air is drawn through an evaporator—consisting of honeycombed cardboard pads moistened with seawater—located in the entry to the greenhouse (see figure 5.9). Once inside, the air (now cooler and more humid) serves as a perfect environment for growing high-value crops; because the arrangement reduces transpiration as much as tenfold,[137] it also reduces freshwater demand.

Warmed by the sun, the interior air is capable of carrying even more moisture. As this extremely humid air is blown across pipes that are filled with cold seawater, it condenses, and the resulting distilled water is collected to support the plants. The air expelled as exhaust from the greenhouse is passed over external vertical evaporators, which are arranged to create a sheltered, humid environment adjacent to the greenhouse. In this area, external evaporative "hedges" (which themselves provide cooling of up to 10°C (18°F) also provide food (cereal crops) and fodder (desert grasses) that help revegetate the surrounding landscape.[138] Finally, useful minerals—including salt and boron, the latter used in glass making—are extracted from the seawater brine (the result of multiple evaporations), thereby adding to the value chain.

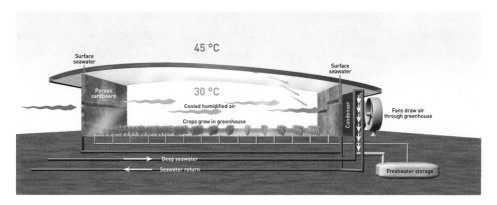

Figure 5.9
Operational diagram of the seawater greenhouse. © Seawater Greenhouse Ltd.

In this arrangement, where cooling and solar distillation are integrated within one structure, waste from each process serves as input for the next. In a pilot project that opened in Qatar in December 2012, the SWG achieved interior temperatures that were up to 15°C (27°F) cooler in summer while using half as much water as comparably sized conventional greenhouses in the area. It also produced bountiful crops—for example, 75 kg of cucumbers per m² (10.76 ft²) annually.[139] The interior environments in these greenhouses have also proven ideal for delicate crops, such as tomatoes and melons. And since saline water controls airborne pests, no petroleum-based pesticides are required.[140] In addition to being inexpensive to operate, the greenhouse is readily constructed using local labor and low-cost materials. It is essentially an extended version of an off-the-shelf, wide-span polytunnel.[141]

Strategically colocating an SWG with a concentrated solar power (CSP) plant can yield even greater synergies. At the Jordan Centre on the Red Sea, near Aqaba, the combination of a power plant and greenhouse will constitute the first phase of the Sahara Forest Project, a privately funded undertaking being led by a Norway-based environmental technology group. At this full-sized commercial facility, known as the Launch Station, multiple energy-, water-, and food-producing technologies are being integrated at scale to create a for-profit enterprise. Consisting of 4 ha (9.8 acres) of seawater greenhouses, 16 ha (almost 40 acres) of outdoor vegetation areas, a CSP plant, and evaporative ponds for salt production, the project will also serve as a platform for research and training.

A CSP is equipped with mirrors that focus light into a heat collector, producing steam that drives a turbine generator. Excess water from the SWGs will perform a cooling function in the power-generation cycle and keep the mirrors clean. Humid air extracted from the greenhouses—along with nearby vegetation made possible by the greenhouses—will "scrub" the air, removing the dust that would otherwise obscure the plant's mirrors. The plant, in turn, will meet the greenhouses' modest electricity needs, with the rest being sent to the grid.[142] In 2015, the European Union committed €750,000 ($819,900) to realize this green energy project.[143] If the combination of a power plant and greenhouse is successful, Aqaba authorities will provide another 200 ha (nearly 500 acres) for expansion.[144]

Other integrated enterprises being planned include the production of algae in waste salt ponds (created from the remaining brine) for use as biofuel or fish food. With input from the U.S. Department of Energy's Algae Biomass Consortium, a test site for this approach has been developed.[145] In a separate project, halophytes—salt-tolerant species that can be used as fodder or bioenergy feedstock—were cultivated in saltwater.[146] Finally, mariculture could easily be associated with the SWGs.

In one of the world's hottest, most arid regions, Charlie Paton's elegantly networked system has the potential to provide power and freshwater, support crop cultivation and reforestation, and foster new value chains from minerals (including salt), nutrients, and algae. Paton notes that similar complexes could even be located away from the coast, preferably at the base of a mountain, where the humid air emitted by the SWG would evaporate and, as it is blown up the mountain, potentially turn to dew or rain.[147]

Looking Ahead: Climate and Infrastructural Ecologies

With the bulk of humanity concentrated along or near coasts, on just 10 percent of the earth's land surface, both "hard" and "soft" infrastructural adaptation to cyclones and tropical storms—both of which are increasing in intensity and frequency—is critical to avert internal displacement, to protect the assets of the poor, and to safeguard water resources while reducing damage to agriculture. Adaptation measures include seawalls, sea dikes, storm surge barriers, wetlands restoration, and floating agricultural systems. As intensive urbanization in poorer countries exacerbates warming temperatures, inland populations as well will be vulnerable to heat waves and the associated health hazards, which will be compounded by increasing air pollution. They will also be susceptible to inland flash floods caused by deforestation and building in floodplains.

As economies in the Global South expand, the demand for water resources will rise accordingly. At the same time, shifts in rainfall patterns, triggered by climate change, will exacerbate existing water scarcity and stress. Unlike "energy alternatives," water has no substitute. But sourcing can be better managed by taking advantage of the hydrologic cycle—storing water when an excess is present (as in Peru and the Fergana Valley) or reusing grey- or blackwater (as in Jordan). In regions subject to desertification, holistic management of soil and livestock can help replenish groundwater. Desalination can also be used to "accelerate" the hydrologic cycle by constructed means. Along with integrated water management regimes, these are some of the pathways to water security that are open to creative nations. In the twenty-first century, economic, social, and political stability may depend on such innovative measures. Although donor funding and carbon markets are providing some financing of climate-smart infrastructure in emerging economies, it is inadequate by a large margin. Thus, the burden falls heavily on the already overextended nations of the Global South.

6

Infrastructural Coproduction: Inclusionary and Participatory Development

Our defining gift as humans is our power to choose, including our power to choose our collective future. It is a gift that comes with a corresponding moral responsibility to use that power in ways that work to the benefit of all people and the whole of life.

—David Korten, *Agenda for a New Economy*

Each year, high in the Andes, members of an Andean community participate in a sacred ritual passed down over the centuries: rebuilding the last remaining rope bridge in the Andean highlands. Spanning deep gorges at great heights, these structures—made of woven *coya*, an abundant hillside nonfood grass[1]—once linked the vast Inca empire, supporting pack animals, runners bearing messages, and imperial armies. After the arrival of the conquistadors, they carried Spanish horses.

The organizational hegemony and political unity of the Inca empire depended on the passage of information and resources across a unique network of roads, trails, and bridges that extended for 39,911 km (about 24,800 mi) along the length of western South America from Lima to Cuzco. Works of innovative engineering, the bridges were the unique handiwork of a nonindustrial but administratively exacting culture that taxed its citizens by requiring them to cooperatively fashion these structures of surprising strength and beauty using scant local resources (and creating zero environmental impact).

During the dry season, many rivers could be forded or be crossed by log bridges, but the turbulence of the wet months called for other solutions. The crossings spanned by suspension bridges could be four times the length of their earth-bound counterparts. Travel accounts claim that the bridges linked between 75 and more than 100 sites in the central Andes.[2]

The vigor of these textile constructions seems disproportionate to the intrinsic strength of their principal component: plant material. To construct the bridges, the

Incas slung massive cables, as thick as a man's body, over the crossing, wrapping them around large stone anchors on either side.[3] The lower cables supported a roadbed made of sticks, lashed together and covered by woven grass mats, which stabilized the U-shaped structure. At waist height, two other cables formed handrails, which were connected to the floor by sidewalls made of thin rope cables.

The Incas' widely dispersed population was unified by a bureaucratic system that relied on public services rendered in remote communities. The institutionalized bridge renewal helped keep the far-flung empire in tow. Local skills were used to erect, maintain, and monitor traffic on the bridge network, which spanned thousands of miles. The annual rebuilding—considered a public service (*mita*)—was the responsibility of a special caste, the Acos people, from whom the *chaca camayoc*, the master bridge builder and designated guardian for each bridge, was selected.[4] Ritualized annual maintenance called for the reconstruction of the entire bridge, through the contributed labor of the nearest village as well as by workers from neighboring communities.

The Keshwa Chaca (straw bridge), suspended about 3,660 m (12,000 ft) above the Apurimac River in the Peruvian Andes (figure 6.1), is still rewoven each year in a civic

Figure 6.1
Woman crossing the Keshwa Chaca (Bridge of Straw), Canas Province, Quehue District, Peru. Photo Credit: Ian Brookshire.

Figure 6.2
Native bridge builders gather during the annual renewal of the last Inca grass-rope bridge, Canas Province, Quehue District, Peru. Photo Credit: Courtesy of Rutahsa Adventures (http://www .rutahsa.com), uploaded with permission by user Leonard G. at https://en.wikipedia.org.

ceremony enacted by the extended community (figure 6.2) and overseen by the bridge guardian. As part of their tribute, women and children reap the grass beforehand. The tubular plant must first be moistened and then hammered flat, worked for pliability, and finally twisted or spun into thin ropes. These are braided by the women into small cables, totaling about 6,700 m (22,000 ft) in length, which the men ultimately twist and braid into several large cables.[5] Stretching the new bridge cables across the gorge takes about 14 hours. During this highly celebratory event, community members share food and drink. This self-organized, obligatory gathering invests the infrastructure itself with social value and reaffirms cultural identity, recalling the community's place in the greater Inca empire.

The Peruvian ritual of rebuilding bridges is but one example of a phenomenon that has been the focus of increasing interest in recent years: community engagement in the delivery of critical services. Discussions of such participation first arose in the 1950s out of a concern that top-down development assistance had failed to adequately consider either the needs or the capacities of the poor and marginalized. Since then, the notion of "inclusive" or "participatory" development has gradually become orthodoxy within the development community, regardless of whether it is successfully or meaningfully implemented. Between 2003 and 2013, for example, the World Bank alone allocated approximately $85 billion in loans for participatory development. Other development entities have made similarly focused contributions, as have the governments of most developing countries.[6]

Inclusive engagement in project selection, planning, and implementation has numerous benefits. To begin with, participation often yields better outcomes, both in terms of improved coordination during planning and implementation and in higher-quality services. Second, participation offers what might be called "social safeguards." It can increase accountability for meeting social targets, help to ensure equitable distribution of benefits, and improve the chances of avoiding corruption. Third, local knowledge can foster the development of culturally and environmentally appropriate technology and facilities. Appropriate technology itself, as evidenced by many of the projects in this book, may be defined by being both "labor- and knowledge-intensive rather than capital-, resource-, and energy-intensive."[7] Finally, participatory transactions, done well, can be socially transformative. Among their benefits are increased employment, civic capacity, greater government-community cohesion, empowerment among women and other marginalized groups, and poverty alleviation, whether through increased equity or the creation of new enterprises.

The "ladder of participation" is often used to describe levels of community involvement in decision making.[8] The lowest rungs represent "induced" participation—potentially manipulative or token forms of engagement in which participants are convened to sanction or otherwise facilitate acceptance of decisions largely made externally. For example, in an approach sometimes known as "public consultation," stakeholders may be asked their views or offered modest choices but will have only a limited share in decision making. Higher levels of participation—ascending from community based, to community partnered, to community driven—can be categorized as more "organic" in their origins and implementation. Roughly speaking, the cases in this chapter are ordered to reflect increasing levels of participation—and to demonstrate the importance of strengthening capacity for shared or controlling roles in long-term infrastructure development.

Decentralization and Community-Based Participation: Moving beyond Tokenism

In the developing world, infrastructure development typically occurs in one of four ways: (1) top-down, government-led (central or local authority); (2) instigated by the private sector (e.g., electric utilities); (3) conceived and/or enabled by nongovernmental or civic organizations; or (4) bottom-up (led by a community or local group). In the first case in this section, the highly centralized government of Viet Nam opted to experiment with decentralization, and even though the initial impetus could be described as top-down, the outcome was democratically empowering. In the next two cases, private electrical utilities in Colombia and Morocco developed new mechanisms for infrastructure implementation, effecting cooperative decision making through formal input—and even in-kind labor—from beneficiaries.

The final case in this section, which centers on an industrial city in India, considers participatory processes for cross-sector infrastructural improvements. As will be further detailed in chapter 8, ensuring that multiple objectives can be met through a single investment demands a concerted impetus (either internal or external) in order to succeed. Such integrated planning is rarely practiced in either the developed or the developing world. Nevertheless, as the case reveals, new delivery models, implemented through broad collaboration, can deliver effective, demand-driven solutions that offer greater productivity and increased benefits.

Decentralizing Infrastructure Development in Rural Viet Nam

Given that infrastructure is both a public good and a source of economic stimulus, governments have historically played a lead role in infrastructure funding and implementation. And, particularly in developing economies, governmental entities may be better positioned to manage the complexity of infrastructural assets, which typically have high costs and long development cycles, and are associated with regulatory and siting issues. Nevertheless, by the mid-1980s, many large-scale, government-led infrastructure initiatives in developing countries had been found wanting when it came to addressing local needs.[9]

Under decentralization, higher levels of government delegate authority to deliver and maintain services to lower levels of government. Ideally, decentralization is accompanied by efforts to build local capacity; in some cases, it can also be an opportunity to improve citizen participation in the workings of local government.[10]

In Viet Nam, the highly centralized, single-party government formally espouses communism yet, in order to gain full entry into the world economy, it has eagerly embraced economic and political reform. Overcoming its aversion to power sharing, the government has successfully delegated responsibility for development to impoverished rural regions, where the projects are answerable to the will of local beneficiaries.

In the 1980s, Viet Nam's two million poorest citizens lived in remote regions that lacked infrastructure, links to markets, and other services; in an attempt to foster socio-economic development in these areas, the government implemented Program 135—a set of decrees and regulations that devolved infrastructural planning and implementation to the commune level (the lowest level of administration in Viet Nam). The projects failed to meet community needs, however; the government also realized that poor communes lacked project management capacity.

In the 2000s, despite a long tradition of centralized decision making, the Vietnamese government made a strategic decision to genuinely empower poor rural communities. In doing so, it followed a model that had been developed in the 1990s by the IFAD (International Fund for Agricultural Development) and further built on by the World Bank beginning in 2000. The goal of the effort was to build capacity by creating opportunities for grassroots, participatory development of small-scale, community-based infrastructure.[11] Despite continued skepticism regarding local management capacity, the central government established standards and processes to guide local actors, and the Ministry of Planning and Investment, in Hanoi, supplied staff to provide the extensive training and capacity building that was integral to the decentralization policy.[12]

Each of the communes received infrastructure grants of between $20,000 and $60,000. It was left to the "project owners"—Commune Project Coordinating Committees (CPCCs), composed of elected village-level representatives—to manage the selection of projects from a given menu of infrastructure options. Communities could prioritize their investment needs, thereby increasing their sense of ownership and improving their bonds with local authorities.

The CPCCs set up consultations at the village level that were supported by trained community facilitators. Relying on villagewide input, the CPCCs undertook design, procurement, and construction oversight.[13] To generate job opportunities, the committees were empowered to hire local labor. The jobs created safety nets for the poor during agricultural downtimes. The CPCCs were also responsible for public information campaigns to gain and sustain widespread support.[14]

Eventually, 760 communes in 93 districts received project support totaling $108 million. This sum funded 8,462 works at the village, commune, and intercommune levels. As a requirement for receiving funds, communities had to meet 5 percent of the construction costs by using either local funds or in-kind labor. The government allocated $9 million for capacity building, and more than 75,000 citizens received training; of those, 2,500 were project staff at all levels.[15]

Overall, the project not only helped develop ownership of the infrastructure (the quality of which was subsequently evaluated as 95 percent satisfactory) through a participatory approach but also helped engender grassroots democracy. Assessment data showed that, on average, 92 percent of households and 90 percent of ethnic minorities participated. To break ties on project selection, women's votes were accorded more weight—increasing both their self-esteem and their sense of responsibility for outcomes.[16] Overall, the Vietnamese government—at both the local and ministerial levels—was pleased with the program's innovative and effective approach. A second phase of the project was subsequently funded by the World Bank to bring decentralization into the mainstream of rural development.[17]

Decentralizing infrastructure planning and delivery democratizes services and is an important step forward for many emerging economies. In the case of rural Viet Nam, it has helped the government reduce poverty while strengthening local decision-making capacity.

Community-Based Shared Use in Colombia's Wayuu Territory

More than 1.3 billion people worldwide suffer from energy poverty—a serious impediment to social and economic development. To provide energy to burgeoning cities and outlying regions alike, many developing nations, particularly those blessed with lakes and rivers, have accepted the environmental and social risks and impacts associated with hydropower dams. Impelled by growing energy demand, other emerging economies are opening up remote, resource-rich regions to other energy infrastructure: signature traces of oil and gas pipelines, roads, and coal-mine excavations now crisscross otherwise uninhabited landscapes.

In the absence of protective measures, much social and environmental harm continues to be done as a result of these activities. In recent decades, however, the development community—along with the investment sector—has established standards and guidance documents designed to reduce or avoid risk both to inhabitants and to the environment on which they depend. Embedded in these measures is the recognition that stakeholder engagement can reveal and mitigate project risk while delivering local benefits.[18]

In the case of large, spatially diffuse, and intrusive infrastructure assets, siting can raise issues of basic self-determination and the right to control infrastructural initiatives that physically impinge on a traditional domain. Approaches may range from the exclusionary (resulting in contested impositions) to consultation, negotiated accommodations, and mutually beneficial colocation, as was the case among the Wayuu—an indigenous people who reside on the remote Guajira Peninsula along the northern coast of Colombia.[19] With a population of about 277,000, the Wayuu—who were never subjugated by the Spanish—represent approximately 20 percent of Colombia's Amerindian population. Their windswept, arid landscape—home to El Cerrejón, the continent's largest open-pit coal mine—supports grazing, fishing, and modest agriculture; women earn their livelihoods through traditional textile weaving. About two-thirds of the Wayuu live in poverty, and the communities lack basic infrastructure.[20]

Colombia's national electric grid is 67 percent hydro based—resulting in a relatively low carbon footprint but leaving the nation reliant on gas-powered backup generation during its two dry seasons.[21] With wind velocities averaging 9.5 km/s (5.9 mi/s), the Guajira Peninsula has ample potential for wind power sufficient to meet dry-season demand even during the peak hours of early evening.[22] Given this natural complementarity between wind and hydropower, as well as the capacity of renewables to reduce exposure to volatile fuel prices, the Jepírachi project—a pilot wind farm—was a compelling investment for EPM (Empresas Públicas de Medellín), one of Colombia's largest utilities.

Under an "easement-of-way" agreement negotiated with the local Wayuu community, the project's 60 m (196.9 ft), 1.3 MW turbines, 15 in all, stand on a 4.9 ha (12.1 acre) coastal site within a 160 ha (365 acre) protected area (figure 6.3).[23] Built between 2002 and 2003, the project has a rated capacity of 19.5 MW and produces approximately 82,000 MWh of electricity.[24] The $27.8 million project cost was financed through a combination of EPM's own resources and loans from the Colombian banking system.[25] Because the project was classified as a Clean Development Project, it received additional financing from the Prototype Carbon Fund, which was created and is managed by the World Bank's Carbon Finance Unit. The fund authorized the certification and sale, in the formal carbon market, of the emission reductions from the wind farm, which are expected to be 430,000 metric tons (473,993 tons) from 2003 until 2019. Accordingly, EPM receives $3.50 per equivalent ton of CO_2 emissions reduced.

The Jepírachi project exemplifies a respectable degree of community participation. EPM engaged the nearby Wayuu community in an extensive consultation and negotiation process that lasted over three years, at the end of which the community authorized the use of the territory (figure 6.3). (The integrity of the process was verified by independent authorities.)

Communicating through community-proposed translators, community-authorized tribal leaders worked with officials from a range of governmental levels (national-level ministries of justice and the interior, the Office of Indian Reservation, and others) to negotiate and ultimately validate the final agreement.[26] Broadly, that agreement spelled out intercultural criteria for shared use of the areas occupied by the wind turbines, power transmission lines, and roads, ensuring that there would be no restrictions on

Figure 6.3
Wayuu women before a Jepírachi wind turbine, Guajira Department, Colombia. Photo Credit: NORDEX.

the Wayuu's ongoing habitation and land use. The consultation and negotiation process focused on technical details, including environmental and social impacts and the management measures that would be undertaken to address them, along with a community benefits arrangement that would improve quality of life, add jobs, and strengthen community organizations.

According to EPM, the environmental management plan, encompassing physical and biotic components, was designed and executed to be "responsive to the socioeconomic and cultural complexities of the Wayuu community in the wind farm's area of influence."[27] To safeguard the traditional agriculture, fishing, and goat herding on which the community depends, the plan included detailed measures to mitigate adverse environmental effects that the community had identified (e.g., noise; interference with daily activities, sacred sites, or rituals), along with specific community-identified protection measures for natural systems. The plan also delineated the bounds of protected settlements, reservoirs, and heritage sites.

A jointly monitored social management plan addressed cultural relations and new, community-driven development programs—improvements in health care, basic sanitation, waterworks (in the form of a wind-powered desalination plant), electrification and distribution infrastructure, and fishing infrastructure. Also included were community benefits designed to support ecotourism, job creation (e.g., training in waterworks maintenance), and microenterprises (handicrafts). According to EPM, handicraft and weaving collectives have strengthened cultural identity, improving social "motivation and cohesion."[28]

Dedicated funds for these community-driven projects—$.50 per ton of CO_2 reduced—flowed through EPM and were directed to the Anna Watta Kai (Well-being for the Future) Foundation, EPM's main institutional counterpart on the Wayuu side.[29] This use of funds was the first project undertaken by the World Bank's Carbon Community Fund, which uses revenue from Clean Development Mechanism projects to support low-income communities.[30]

EPM claims that since the company undertook the project, respect for indigenous customs and the use of best practices for trust building, ongoing community participation, and interinstitutional management have become embedded in its corporate social responsibility policy. With other wind resources still largely unexploited, a number of international entities have proposed similar installations on the peninsula.[31] One under consideration—a 20 MW site at nearby Joutkai—would generate sufficient revenue to support a rural service utility providing basic electricity and potable water to 90,000 inhabitants. (As of this writing, the project had not advanced.)[32]

Despite the positive experience of the Wayuu at Jepírachi, many indigenous inhabitants still remain wary of the incursion of large corporations, especially given the social and environmental disruption of the coal mine.[33] Nevertheless, this low-carbon energy project offers a good example of partnering with affected communities to mitigate social and environmental impacts, strengthen community-based participation, and achieve community benefits.

Energy Access for the Marginalized: A Minigrid in Casablanca, Morocco

Utility providers working in emerging markets rarely see solving energy poverty as part of their core business strategy. This is especially true for the informal settlements inhabited by the urban poor, many of whom lack secure land tenure or even permanent addresses. And yet, for the world's 1.3 billion inhabitants who lack electricity—and the nearly double that number who lack clean cooking energy—basic health and economic development are at stake.

Among the poor who live near the grid, many engage in "energy theft" (illegal power connections); the rest rely on fuels like kerosene for lighting and traditional biomass burning for food preparation. Trapped in energy poverty, the bottom 20 percent of households spend an exorbitant amount of their income on these unhealthy fuels.[34]

In emerging economies, electrification has proven problematic for both customers and concessionaires. Connections are typically out of reach for the urban poor, and, since many informal settlements exist under threat of eviction or demolition, the population tends to be highly mobile and thus regarded as a poor investment for utilities. For those at the bottom of the pyramid, the energy market has been left fragmented and underdeveloped. Some multinationals and local private companies, however, have been investigating novel initiatives designed to work in partnership with citizen organizations.[35]

In Morocco, despite efforts to curb theft, the Lyonnaise des Eaux Casablanca (LYDEC)—a private consortium that services the city with power, water, and sanitation—found that more than 150,000 households in 130 settlements were tapping electricity illegally. The resulting expenses and losses amounted to more than €1.4 million annually, a gap LYDEC sought to close by persuading the city government to allow service to be extended to several of these vulnerable informal settlements, specifically those that might under government policy be either upgraded or eliminated at some much later date.[36]

The Temporary Lower Cost Mini-Grid Electrification program, established in 2004, created temporary minigrids to serve groups of householders in these settlements, linking them through transformers and master collective meters that were sited in areas where they could be readily removed on short notice should the municipal authorities decide to eliminate the settlement. The minigrids were operated and maintained by local community representatives elected by each settlement group. In addition to serving as trusted intermediaries between the utility and the community, these representatives were given technical and managerial training so that they could connect a secondary network, consisting of about 20 additional households, in a "block."[37] The program was designed so that each household paid the representatives according to its sub-meter reading. If the amount owed on the entire minigrid was not paid in full, LYDEC could disconnect that circuit. Thus, peer pressure was enlisted to help with revenue collection.

The results were notable: 93 percent of the households in the participating communities were legally electrified, enjoying lighting, refrigerators, and satellite TV, at a single connection cost of $130. LYDEC received 98 percent of what was billed, allowing the project to recover its initial costs within three years.[38] Within five years, 75 percent of the households in Casablanca's informal settlements (30,000 households) had been connected. Eliminating its billing and collection costs worked to LYDEC's advantage, enabling it to meet the needs of the settlements while maintaining profitability. Businesses within the settlements increased by 17 percent, and 3,500 new jobs were created—including the 1,250 community representatives.[39]

The breakthrough behind this social-marketing initiative was "coproduction"—the delegation of key elements of service delivery and management to individuals drawn from the communities. This aspect of the arrangement reduced LYDEC's investment, strengthened its relationship with the settlements, and improved social cohesion within the communities.

The program—whose effective, efficient approach to energy access has been successfully replicated in Tangier-Tetouan and Rabat[40]—demonstrates how innovative community participation can overcome barriers and help motivate utilities' investments in electricity access for the poor. Although the project does not address carbon reduction, it establishes a template for powering rural areas through the establishment of a collective utility, regardless of the power source.

Cross-Sector Coproduction in Nashik, India

Can interlinked and interdependent infrastructure, strategically implemented through joint planning and participatory approaches, add value? The Urban NEXUS program—framed by ICLEI South Asia and the German Development Cooperation under the leadership of Germany's Federal Ministry for Economic Cooperation and Development (BMZ)—provides one answer to this question. The ministry's implementing agency, the Deutsche Gesellschaft für Internationale Zusammenarbeit GmbH (GIZ, or German Society for International Cooperation), has undertaken pilot projects throughout the world. One such project is in Nashik, a large, industrialized city in central Maharashtra, India, with a 2011 population of almost 1.5 million.[41]

In addition to being an industrial leader, Nashik—known as both the "backyard of Mumbai" and the "wine capital of India"—is a major locus of urban and periurban agriculture.[42] Farming contributes a significant portion of the city's GDP—hence the importance of the Urban NEXUS project, whose objective was to "identify and establish mutually beneficial responses from the inter-linkages between [the] sectors [of] water, energy, [and] food."[43]

Across India, irrigation water supplied by canals is supplemented with groundwater withdrawal. This requires energy-intensive pumping, and in order to boost the nation's capacity to feed itself, federal policy has long provided substantial electricity subsidies. Meanwhile, both the overexploitation of groundwater and the lack of accountability for pumping have led to water overuse and energy waste. Moreover, in a situation where water and energy are mutually dependent, the agriculture sector is completely reliant on the unsustainable resources of these two sectors.[44]

In Nashik, which has almost 15,000 ha (37,065 acres) in cultivation within its jurisdiction, the groundwater table has fallen by 1–3 m every five years for the past several decades.[45] But typical barriers had made solutions elusive. Resource planning fell into silos: ten separate municipal and state departments. Political interference, bureaucratically compartmentalized decision making, lack of regulation and enforcement, and limited interdepartmental communication had intensified the disconnect across the food, water, and energy sectors—which, in Nashik alone, made up three separate city departments: land use (food production), potable water supply (City Corporation), and a separate irrigation department.[46]

GIZ and ICLEI regarded the city as a perfect candidate for an Urban NEXUS pilot project and designated the Nashik Municipal Corporation (NMC) as the lead actor. In early 2014, a dedicated 30-person stakeholder team was assembled to initiate a dialogue. Members included staff from the city's water, sewerage, and planning departments; staff from district- and state-level irrigation, agriculture, and electricity-distribution departments; and representatives of environmental engineering and agricultural training programs from various academies.[47] Uniting its cross-sector knowledge and capabilities under this common platform, the stakeholder group was able to pinpoint key vulnerabilities and systemic problems afflicting Nashik's agriculture. Recognizing that improving resource efficiency would call for inclusive, cooperative decision making and increased flexibility across institutional cultures,[48] the team selected four strategies with synergistic potential to relieve the stresses of the food-water-energy nexus.

The first involved waterworks that had been installed by local farmers—some legally registered but others not. Collectively, these consumed as much as 20 percent of Nashik's energy production. Of the 35.7 million registered agricultural pumps, most had operating inefficiencies of up to 20 percent, largely because of improper sizing, poor installation, and inadequate maintenance. Moreover, because the energy used for pumping was subsidized and unmetered, farmers were not accountable for the amount they used.[49] Finally, rainwater harvesting had been neglected as a means of groundwater recharge, leaving many areas critically short of vital groundwater.

Farm community presentations were conducted by the NEXUS team to gain buy-in from the farmers, to build their capacity to help install and operate the efficiency improvements, and to obtain their agreement to provide ongoing compliance audits. Community ownership was confirmed through tripartite agreements between the farmers, NMC, and the NEXUS team leadership. Under these agreements, low-friction pumping and right-sized pumps were installed in the pilot districts, along with drip-irrigation and underground supply pipelines and metering devices. These interventions reduced energy use by about 20 percent.[50]

Since groundwater had been supplying 70 percent of the water for crop irrigation, NMC (along with the state and district Groundwater Survey and Development Agency) suggested a second intervention, which involved augmenting the flow of surface water into subgrade reservoirs to create groundwater recharge systems in zones that NMC had identified. Maintaining the water level at a more easily reached depth would then synergistically reduce the pumping energy required. NMC subsequently undertook a study to explore a citywide groundwater recharging system. Although NMC has allocated €73,200 from its budget to implement groundwater storage in the city, outside technical guidance and capacity building will be needed to scale up the program.[51]

Like the first initiative, the third also focused on integrating the farming community into development efforts. To reduce fossil fuel consumption while creating a renewable energy supply system, the Urban NEXUS stakeholder group worked to revive the use of biogas in the pilot area. In the 1980s, under India's National Biogas and Manure Management Program, local farmers had installed biogas plants; however, as mechanized farming and urbanization had reduced local livestock, leading to declines in biomass and manure input, the plants had fallen into disuse, and the community had resorted to wood or other biomass combustion to meet energy requirements.

When the stakeholder group mapped biogas potential throughout the city, it found that a combination of poor maintenance and lack of raw material had rendered 52 percent of the plants nonfunctional. They then used federal financial assistance, channeled through the state governments, to install 650,000 new biogas plants and repair nonfunctioning ones.

As its fourth intervention, the Urban NEXUS team fostered more routine interactions between farmers and local citizens and developed educational programs for children and farmers. The purpose of these efforts was to generate ownership of the project and foster the behavioral changes needed to help Nashik meet its environmental goals. One such initiative brings students to model farms so that they can better understand the connections between sectors, view various approaches to water conservation, and receive an introduction to organic farming.[52]

Through a facilitated process, the Urban NEXUS program gave NMC officials a powerful taste of collective decision making. Among its other benefits, the pilot project helped instigate Nashik's Sustainable Urban Habitat Action Plan, an integrated, cross-sector initiative to address climate change mitigation and adaptation.[53]

Partnering for Service Provision

As the preceding cases demonstrate, once utility companies, local governments, and central governments come to understand the value of community input to infrastructural design and operations, they move beyond the lowest levels of participation. The cases in this section illustrate genuine coproduction, whereby a community contributes resources (knowledge and labor) to achieve service delivery. Such actions are in the realm of organic participation, where communities act more autonomously and sometimes even in opposition to central authorities. Motivations can range from the instrumental, as in the case of coastline rebuilding in Viet Nam, to identity-based or ideological, as in the case of the Brazilian waste pickers who sought recognition of their services.

Ideally, community involvement in government- or donor-led (NGO) projects will yield infrastructure services that are more aligned with the needs and preferences of the customers and more likely to be well maintained. At their best, projects that require direct community participation in construction or maintenance, or some form of cofinancing, engender a greater sense of commitment and ownership. The success of such efforts ultimately depends, however, on local capacity: the community's ability to initiate and follow through on collective action.

Community-Based Management of Coastline Resilience: A Participatory Process in Viet Nam

As noted in chapter 5, Viet Nam recognizes the urgency of protecting its coastline against sea-level rise and storm surge. In the Red River Delta of northern Viet Nam, a successful initiative sponsored by the Viet Nam Red Cross had fostered coastal resilience by rebuilding dikes and planting mangrove forests. In the wake of Typhoon Damarey—which in 2005 ripped through two miles of dikes, inundating six villages and salinating agricultural fields, fruit orchards, and cattle farms—similar fortifications were needed in the remote coastal commune of Da Loc, which is situated south of Hanoi. In 2006, at the instigation of CARE, an international nongovernmental organization, Da Loc's mangrove forests were regenerated through an integrated approach that included training for community-based maintenance and protection of the restored ecosystems. Along Da Loc's rapidly established mangrove strip, which is nearly 3.21 km (2 mi) long, the community planted more than 2,000 plants per hectare (809 per acre), protecting more than 6,000 people.[54] Not only did CARE's approach strengthen the community's embrace of adaptation strategies, it also ensured even the poorest and most vulnerable inhabitants legal access to more equitably managed commons resources.[55]

Mangrove wetlands are typically lost through conversion to other uses, principally agriculture and shrimp aquaculture. Two factors have reinforced this pattern: first, the ecosystem services provided by these rich wetlands have been undervalued; and second, these coastal barriers have historically been treated as a public good and therefore subject to nonexclusive use.[56] CARE understood that fostering community-based natural resource management was the key to both communal risk management and the protection of a common-pool resource.

In a nation where democratic participation was not yet common, CARE established the Community-Based Mangrove Management Board (CMMB)—a democratically elected, 20-member group made up of farmers, appointed village officials, and representatives of civil society organizations (e.g., the Women's Union).[57] Created with the approval of the Thanh Hoa Provincial People's Committee (the executive arm of the provincial government), the CMMB had authority and responsibility for arranging technical training courses and organizing participating community members into work groups:[58] the "nursery group," the seedlings "planting groups," and the mangrove plantation "cleaning and protection groups" (see figure 6.4).

The villagers, newly responsible for the protection of their own land, contributed valuable ideas for planting and maintaining the mangroves; for example, knowing that barnacles and small shells killed the trees, they removed the barnacles and shells by hand. The survival rate of the trees has been high: about 80 to 85 percent. Moreover, as the capacity of the CMMB has grown (it has entirely led the process since 2009), the cost of planting the mangroves has decreased. To improve rice production, the CMMB also undertook the regeneration of salt-afflicted lands, developing, with assistance from CARE, a series of canals to flush the contaminated lands with river water, which more than doubled yields.[59]

To support the development of a participatory forest land-use plan, CARE also provided the CMMB with training in legal rights and responsibilities. As a first step in plan implementation, CARE helped the community conduct a participatory inventory of local ecosystem resources. Using GPS technology, villagers estimated the resources that

Figure 6.4
Workers cleaning the beach near Da Loc, Viet Nam, during mangrove restoration. Photo Credit:
© Søren Rud.

would be available for sustainable harvests, in five-year increments, to begin to establish harvesting quotas. Under the plan's regulatory policies on mangrove use, which
cover the exploitation of forest resources for firewood, seeds, beekeeping, and seafood
harvesting, users must also contribute financially to mangrove protection and management. Villagers can also contract with the CMMB to lease land, enabling them to profit
from the forest while safeguarding its resilience. Under the new system, approximately
3,000 people now have opportunities and rights to sustainably utilize the forests—
harvesting fish, crabs, and mollusks to augment their incomes. In addition, the poorest
households are allowed to collect dead forest wood for use in cooking.

To improve ecosystem awareness and build capacity in the villages, CARE worked
with youths to create Green Teams—elected student representatives and student unions
that clean the beaches, maintain the mangroves, and encourage recycling and composting in order to produce organic fertilizers that can displace petroleum-based substances.

As a means of ensuring participation and a confident sense of ownership, as well
as access to project-related benefits, the CMMB's Women's Union offered dedicated

training courses for women.[60] CARE's experiences in reducing poverty and building capacity, both in Viet Nam and elsewhere, have shown that when women earn income (and control the use of that cash), they can play a central role in lasting solutions.[61] According to the chairwoman of the Yen Loc village women's union in Da Loc, coastal communities' living standards rose with the added income from controlled aquaculture.[62]

After three years, the program was deemed a success, thanks to the participatory approach. The project established formal recognition for the more widespread use of community-based management systems that outline clear community rights, set standards for comanagement, and provide for ongoing, local organizational structures. The provincial government and the district People's Committees have endorsed expansion of the project; meanwhile, the Vietnamese Ministry of Agriculture and Rural Development, which is planning to establish 100,000 ha (274,105 acres) of government-led mangrove planting and management, has set up its own experimental plot in Da Loc to learn from the CARE model.[63] As a strategy for mangrove conservation and restoration, collaborative natural resource management is also beginning to emerge in several other villages.[64]

From Public Nuisance to Public Asset: Coproducing Sanitation in Senegal

In the early 1990s, rising sea levels mobilized nine Senegalese communities in the coastal township of Rufisque to finance, construct, and operate their own waste and sanitation systems. They did so by mastering technologies they could control, relying on democratic decision-making processes, and working in concert with local authorities.

The commune of Rufisque (from the Portuguese *rio fresco*) is made up of low-income villages located to the southwest of Senegal's capital city, Dakar. In the early 1990s, the roads were poor, and the compounds lacked piped water, sanitation, and refuse collection. Household wastewater and garbage were left in the street or on the beach, leading to dysentery and other diarrheal diseases as well as skin infections. Moreover, the coast was subject to severe erosion: over the previous three decades, the beaches had shrunk from 300 m to 20 m (65.6 ft), in part because of sea-level rise and storm surges.[65]

In 1980, working with the local public works department—and with financial aid from France and Finland and technical assistance from ENDA-Third World/RUP (Relais pour le développement Urbain Participé)—the villagers of Diokoul, one of the older villages in Rufisque, had constructed a series of dikes that jutted into the sea. The dikes broke the current, fostered beach nourishment (sand replacement), and secured the town from damaging storm surges. During this endeavor, ENDA-Third World/RUP recognized the organizational and management capacity of Diokoul's citizens, who, the following decade, requested help with hygiene problems that the Dakar municipal government had been unable to address.

In 1991, with community input, ENDA-Third World/RUP drew up plans for the PADE (Programme d'Assainissement de Diokoul et quartiers Environnants), the Diokoul and Surrounding Districts Sanitation Scheme. Agreements followed with a number of entities, including the Canadian Host Country Participation Fund, the Rufisque Local Authority, the Senegalese Water Ministry, and private businesses. Neighborhood committees, including Diokoul's Health Committee, were also parties to the agreements, and helped by serving as social intermediaries. Collectively, the signatories made up the Local Management Committee.[66] With training provided by ENDA-Third World/RUP, the residents became partners and contributors as well as beneficiaries, paying all of the initial costs for the wastewater and sanitation facilities and 70 percent thereafter (some residents' contributions are channeled into a revolving fund); residents also cover maintenance costs and 50 percent of the provision for depreciation.[67]

The household sanitary waste system was designed to be low cost and relies, in part, on a traditional practice that had been used in Sudan for millennia. Domestic water is piped to the households, and wastewater is piped to a "purification plant"—a reservoir where "water lettuce" (an aquatic plant native to Africa) is used to remove sediment and purify the water. The water lettuce's biomass is eventually combined with collected organic waste and the treated water in order to produce compost for agriculture. A system of door-to-door refuse collection, performed by horse-drawn carts, transports garbage to collection points, where it is then transferred by trucks, belonging to the local authority, for incineration.[68] The project beneficiaries jointly established the Fonds Communautaire pour l'Assainissement des quartiers Urbains Pauvres (FOCAUP), the Community Fund for Sanitation in Poor Urban Neighborhoods, a mostly self-sustaining, revolving fund supported by fees from citizens along with subsidies from external partners. Through the Local Management Committee, residents decide how the fund's resources are to be allocated for system expansion. FOCAUP offers credit to households that apply to join the sanitation system. The high repayment rate—73 percent[69]—demonstrates that local credit initiatives can compensate for a lack of government financing.

The new sanitation systems have freed women from some of their daily tasks, allowing them to engage in activities outside the home. Since they make up about 30 percent of the heads of households that benefit from private sanitation, women play a prominent role in the project, participating in decision making and overseeing infrastructure maintenance.[70]

The project's outcomes were significant. The sewerage systems serve more than 550 households, representing 3,850 individuals. More than 8,320 households benefit from garbage precollection by carts, 50,900 metric tons (56,000 tons) of which are siphoned off annually for composting.[71] In addition to reducing the incidence of microbial illnesses, the integrated approach to refuse management and wastewater treatment costs 70 percent less than other municipal treatments. By exploiting the connection between these two hygienic practices and using alternative neighborhood-level technologies, this simple, productive infrastructure also created 35 permanent jobs.

The Diokoul project has inspired similar initiatives in outlying Rufisque and Dakar, and even the Senegalese cities of Yoff, Niaga, and Saint Louis have benefited by adopting the model. Requests for information have come from as far away as Mali and Mauritania. In 1996, Habitat II, a conference on human settlements, selected the Rufisque ENDA-Third World/RUP initiative as a best practice.[72]

Professionalizing the Waste Sector: The Cooperatives of Belo Horizonte, Brazil

In cities around the world, the collection and disposal of urban waste has long been the purview of self-organized but largely unregulated and unprotected enterprises. Rarely do these informal economic ventures gain legitimacy, but their services, as recounted in the discussion of the East Kolkata Wetlands in chapter 3, are critical to overburdened municipal budgets. Whereas in East Kolkata the informal sector relies on natural systems to process much of the waste, in other cities across the globe much of the collecting, sorting, and recycling is undertaken by informal workers, many of them members of vulnerable groups such as migrants, women, children, and the unemployed. The work is dirty and degrading, and the conditions are inhumane; worse, daily contact with waste is fraught with health hazards.

As of 2008, around 15 million people survived by collecting and processing recyclables.[73] These workers have been neither socially nor professionally recognized and are often denounced as vagrants. An exception can be found in Brazil's Belo Horizonte, a municipality cited in chapter 4 for its early adoption of landfill gas capture for energy generation. That effort was one element in a larger initiative designed to bring integrated solid waste management (ISWM) to the city.

In the 1980s, Belo Horizonte's waste pickers existed as an underclass—living on the street, scavenging from trash bags, and strewing debris. The waste pickers' homelessness—and its effect on the public space—ultimately helped mobilize Belo Horizonte's ISWM program to address both social and environmental concerns. Under the ISWM program, the city's waste scavengers—sweepers, pickers, collectors, packers, and sellers—formally partnered with the municipal administration, creating an efficient urban waste management system that professionalized their positions, systematized their processes, and improved their working conditions. In 1990, Belo Horizonte adopted techniques for participatory democracy that fostered citizen involvement in establishing public priorities for budget allocation and laid the groundwork for efforts to improve municipal services while generating income for the poor.[74] In 1993, the Catholic Church assisted in organizing street scavengers, known as *catadores*, to create Associação dos Catadores de Papel, Papelão e Material Reaprovitável (ASMARE), the Association of Collectors of Paper, Cardboard and Recyclable Material. At around the same time that lack of waste management had begun to gain public attention, ASMARE asked the city to officially acknowledge its work in managing recyclables. Recognizing that a formal partnership with ASMARE had the potential to increase overall recycling rates, the city changed local legislation to mandate collection of recyclables by the waste pickers, who were by then organized in several cooperatives under ASMARE.

With ASMARE as its principal partner, the city established the Superintendency of Public Cleansing (SLU) to develop procedures for the selective handling and treatment of the city's solid waste.[75] The SLU's first step was to create a multidisciplinary team with two responsibilities: to mobilize and educate the general population and to train the waste workers in their newly defined roles within a comprehensive system. Next, the SLU established 150 curbside drop-off sites where the public could deposit presorted recyclables. These recyclables were collected and brought to warehouses run by the cooperatives established by ASMARE. Meanwhile, using cooperative-owned vehicles, the cooperatives also collected recyclables from industries, commercial entities, and public offices, dropping them at the waste management warehouses (figure 6.5).[76] At the warehouses, workers further sorted, shredded, baled, packaged, and stored various materials for sale to industry. Many of the new warehouses had mechanical sorting systems suited to the waste pickers' requirements, and to protect the workers' health and safety, the city provided clothing to be worn in the warehouses. The city also formalized the work of another underclass, the *carroceiros*—drivers of horse-drawn carts—who were now paid by households and industry to carry construction debris to designated sites.[77] As of 2012, 95 percent of the city's urban population and 70 percent of its favelas benefited from domestic waste collection.

Today, 600 waste pickers are members of cooperatives that work in 80 warehouses. Their working conditions have significantly improved, and they earn a monthly minimum wage of $321.[78] Workers are also eligible for a monthly 20 percent productivity incentive based on their output.[79] Through their membership in ASMARE, the waste pickers have created a self-supporting business. Since 1997, the waste pickers have been managing all aspects of the work they carry out, including the marketing of recyclables.[80] ASMARE has multiple committees, which handle issues from infrastructure to health, finances, education, and culture. It works with local government through an entity known as the Collegial Board, whose role is to further define and integrate the roles of each of the players.[81]

Figure 6.5
A member of the waste picker cooperative, Belo Horizonte, Brazil. Photo Credit: Courtesy of Sonia
Dias.

As a self-supporting, autonomous community enterprise, ASMARE has diversified: it
now runs a restaurant and has developed a paper-recycling unit. It has also expanded
operations into nearby small towns. Perhaps the most important outcome occurred in
2010, when, as part of the national government's new solid waste management policies, the role of waste pickers' cooperatives was formalized nationwide.[82]

In Belo Horizonte and many other parts of Brazil, landfills remain the primary form
of waste disposal. Nevertheless, the formalization of the waste-scavenging sector has
been widely viewed as socially progressive, environmentally sound, and economically
viable. In other countries (Colombia, India, and Uruguay, for example), governments
have initiated similar programs.[83] With its joint development of action plans and consolidation and legalization of institutions, this case represents participation at one of
the higher rungs of the participation ladder.

Entrepreneurship and Comprehensive Citizen Control

When communities desire services from a government that lacks either the resources or the will, some choose to contribute their own labor, funds, or both in order to meet their needs. In the cases considered in this section, civic groups acted independently of external institutions to implement their own infrastructural systems. Although they may have sought advice and resources from outside entities, they maintained control of the outcomes. Such undertakings—true expressions of communal empowerment—can sometimes establish a pattern of local entrepreneurship and may even catalyze change elsewhere, as in Belo Horizonte.

In the first case, an entrepreneurial, communal agroforestry initiative led to the establishment of infrastructural services in a remote community; in the second, a remote village that the government had left behind undertook an exemplary response to water stress. These examples of community-driven service delivery reveal that bottom-up civic action can force the hand of—and eventually leverage support from—central authorities. More importantly, such endeavors build social capital, create new institutions, and empower new or existing authorities.

Infrastructure, Carbon Mitigation, and Adaptation from a Sociocultural Enterprise: Ixtlan de Juárez, Mexico

Founded in 1988 in one of the poorest regions of northeastern Oaxaca, Mexico, the Ixtlán de Juárez Communal Forest transformed 2,479 impoverished Zapotec farmers into owners of a diversified industrial enterprise organized around a sustainably managed forest commons.[84] Unique governance of common resources has afforded the community electricity, streetlights, water supply, sewage treatment, road and bus service, education, and public health services. More remarkably, those same resources are sequestering carbon, storing valuable water, and providing a pathway for regional climate adaptation.

Until the 1986 passage of a forestry law recognizing communities' rights to form their own logging businesses, a government-sanctioned, state-owned timber-for-paper concession had prevented the indigenous Zapotecs—the legal owners of the forestland—from using their own resources. They were forced to survive within areas that had been cleared for commercial agriculture and cattle rearing.[85] As the paper company intensified its logging activities, wide swaths of woodland were severely damaged. Once they gained socioeconomic control over the forests, the Zapotec—through the Forest, Agriculture and Services Communal Enterprise of Ixtlán de Juárez—established an ecosystem-based management system governed under their own traditions. Under this system, springs and biodiversity habitats are protected to ensure survival of desirable medicinal and ornamental plants and mushrooms, and logging takes the form of small-area clear-cutting, known as *matarza*, which allows the regrowth of many plant and tree species.[86]

Ixtlán's community enterprise encompasses 21,100 ha (52,137 acres) of forests, mostly pine and oak, only 7,648 ha (18,900 acres) of which—36 percent—is under intensive commercial forest management. In 2000, Ixtlán de Juárez won the government's National Merit Award for Sustainable Forestry Management.[87] And its forest enterprises have been certified as sustainably managed by SmartWood, a U.S.-based global forestry certification program sponsored by the Rainforest Alliance.[88]

The community's award-winning forestry program has transformed the forest enterprise into a business incubator, with employment opportunities that would have been unimaginable to previous generations. In addition to caring for the woodlands, community members work in a computerized tree nursery that produces 500,000 seedlings annually; some are designated for Ixtlán de Juárez, and the rest are supplied to other community forest enterprises in the region. Since 2005, a newly constructed industrial park has housed two sawmills, drying ovens, and a high-tech furniture-making factory that produces green-certified doors, desks, and bookshelves for schools and government offices (an exemplary case of vertical integration). Finally, Ixtlán de Juárez has created an ecotourism program that features lodges, restaurants, and other amenities. By providing a high level of local opportunity, the community has achieved one of its expressed goals, which was to reduce the emigration of its youths to cities.[89]

Ixtlán de Juárez's form of community capitalism, which employs approximately 300 workers, is run according to ancient *usos y costumbres* (uses and customs), which are rooted in community service and consensus-based decision making. Obligations include *tequio* (weekly physical labor in service of the community) and *cargos* (serving in an assembly). A community member who has completed his or her *cargos* becomes a *comuneros*—officially part of the "neural network of community knowledge"; a *comuneros* earns wages well above the Mexican average and receives unique social and health benefits.[90]

Governmental environmental agencies and the National Forestry Commission have been engaged with the community as supportive institutions. For example, the state and federal governments have given Ixtlán de Juárez advice and funding for critical infrastructure investments. As part of its effort to expand its economic base through ecotourism, the community has also partnered with the World Wildlife Fund.

Most of the socioeconomic benefits enjoyed by the community, however, are derived from the community-based forest management program. In 2009, the collective businesses made a $230,000 profit, of which 30 percent was reinvested, 30 percent directed to forest preservation, and 40 percent funneled to the workers and community pensions, credit unions, and support for Ixtlán students studying at the state capital.[91] Other benefits include greater opportunities for women, who make up 80 percent of the furniture factory employees and hold leadership positions in the community.[92] An indirect but consciously managed benefit of forestry management is improved water infiltration. Taking advantage of the amplified underground springs, another nearby Zapotec community, Santa María Yavesía, has created a cooperative water-bottling business.[93]

Ixtlán de Juárez's enterprises are also consequential from the perspective of climate change mitigation. In a world where deforestation accounts for 20 percent of carbon emissions, carbon capture through reforestation is an essential strategy (reforestation also reduces erosion, enabling soil to store more carbon).[94] In Ixtlán de Juárez and surrounding community forest enterprises, reforestation relies on nursery plantings of coniferous fir seedlings (figure 6.6). Community forestry management has succeeded in increasing carbon capture in the region: over a two-year period, the Sierra Norte region of Oaxaca showed a 3.3 percent expansion of forest cover while other noncommercial forests suffered high rates of deforestation.[95]

Thanks to the training that young community members have received in silviculture and meteorology, Ixtlán de Juárez is also undertaking adaptation measures that will help prepare its forests, as well as others, to cope with global warming.[96] To preserve the species being threatened by rising temperatures, community workers are prudently relocating the several native pine-tree stocks that are at risk to locations hundreds of meters above their current growing altitudes.[97] The stock can then be sold to other regions for similar adaptation initiatives.

Ixtlán de Juárez's community-driven forest management efforts and commercial enterprises have enabled it to compete in global markets while conserving a natural legacy for future generations. Equally important, the community is a real-world refutation of "the tragedy of the commons."[98] By showing that commonly held property *can* be sustainably managed,[99] Ixtlán de Juárez has also demonstrated that government forest management can be successfully devolved to local communities.[100]

Conservation experts believe that devolution may be gaining momentum. According to *The Economist*, between 75 percent and 80 percent of Mexico's temperate and tropical rainforests are controlled by indigenous populations engaged in community enterprises, placing Mexico ahead of many developing nations with regard to forest management best practices. Since 1990, the extent of forests in developing countries under local community domain has doubled to over 400 million ha (988.4 million acres), or 27 percent of the total.[101] This is a promising development. Given the long history of governmental mismanagement of forest resources (or their co-option by private enterprise), community management is not exempt from collusion and corruption;

Figure 6.6
Pine seedling nursery caretaker, Ixtlán de Juárez, Mexico. Photo Credit: Kumar Sambhav Shrivastava.

nevertheless, from the perspective of social equity and sustainability, even the most flawed community forestry program may be preferable to typical governmental and private sector initiatives.

In Ixtlán de Juárez, social entrepreneurship enabled a once-marginalized sector of society—a community devoid of infrastructure—to experience economic prosperity. Ixtlán de Juárez's achievements in the realm of self-governance further empowered its citizens to gain full access to public services. Ixtlán exemplifies the top rung of the ladder of citizen participation: citizen control.

Water for and by the People: Community-Driven Infrastructure in Lufumbu, Tanzania
In the early 1990s, just half of the widespread population of sub-Saharan Africa had any access to safe drinking water. In the small, isolated community of Lufumbu, Tanzania, the lack of household water connections meant that women had to walk up to 2 km (1.24 mi) while carrying water jars on their heads.[102]

When the government turned down Lufumbu's application for water infrastructure, citing a lack of funds, the disappointed villagers took action. Using predominantly local resources (which included contributing 48 percent of the $50,000 project cost), they constructed an effective water supply scheme that functions under the authority of a democratically elected water committee.[103] At the time the project was undertaken, Lufumbu relied on subsistence agriculture—maize, beans, bananas, cereal crops, and coffee—that produced an average income of about $1 a day. The project served as an effective entry point for other development initiatives.

The infrastructure Lufumbu developed—a gravity-fed storage system that taps a distant source—relies on appropriate technology (that is, it reflects sociocultural, economic, and environmental conditions) and serves the entire community. The system consists of a series of 60,000 l (15,850 gal) tanks that serve as reservoirs and 10 km (6.21 mi) of water mains and distribution pipelines, with 56 drawing points. Tanks are capped with corrugated steel roofing to reduce evaporation.

Assisted by water technicians, the villagers built the infrastructure themselves—in four and a half months—using a self-organized division of labor. The work was broken down into seven segments, each managed by committee.[104] Villagers dug the ditches. Adult men ferried the locally sourced building materials (the villagers chose stones instead of cement blocks, which reduced the cost by 30 percent). Women and children brought the workers food and water. When the project was complete, women and children could walk a mere 200 m (656 ft) to the drawing points—less than the national target of 400 m (1,322 ft). All operations and maintenance are handled by the beneficiaries, many of whom have undergone training.[105] Not surprisingly, the village's innovative, efficient, and transparent governance system created a sense of ownership.

Access to a clean, stable water supply significantly improved living conditions, reducing water-borne diseases by as much as 80 percent;[106] allowing villagers to add new vegetables and staple food crops; and supporting the establishment of coffee as a cash crop. The infrastructure improvements have also reduced women's workloads, allowing them to reallocate their time toward agriculture and other income-producing activities. Thanks to the new water sources, villagers are reforesting their land. Formerly barren hills are replete with new vegetation, improving biodiversity and increasing water retention in the catchment areas. Even housing improved: with water available to make brick, more than 325 new homes were built.[107]

The project legacy includes a town governance system that is fully responsive to community needs. The water management committee not only handles day-to-day water infrastructure issues and establishes regulations and user fees but has begun to integrate sanitation and resource conservation into its purview. In 2004, under the Global Environment Facility (GEF) Small Grants Programme, a United Nations Development Programme (UNDP)-implemented community financial and technical support initiative, Lufumbu received funding for another intake and storage tank, as well as for the provision of public training in health and sanitation.[108]

The significance of the Lufumbu case is its demonstration that infrastructure projects—including those that may be more complex and technical than other development initiatives—can indeed be produced, owned, and operated by poor communities. The district government subsequently adopted the water infrastructure design as standard for all community-based water systems.[109]

Stepping Up the Ladder

In many emerging economies, the inclusion of new actors—the users of infrastructure services—is helping to restructure financing, decision making, and asset governance. It is also accelerating the decentralization, localization, and democratization of service production and operation. Moreover, it represents an emergent prioritization of the customer by public and private utility sectors. Finally, inclusive participation has been instrumental in the colocation of services and other infrastructural ecologies.

Community engagement builds on the often underappreciated skills and knowledge of users, as in Viet Nam's mangrove rehabilitation and Lufumbu's potable water provision. This added capacity helps balance decision-making power between those providing technical expertise and financial support and those who will actually benefit from—and, ideally, take ownership of and operate—the services. Importantly, infrastructural interconnections and interdependencies rely on community trust and cohesion, among the most salient characteristics of developing economies, where community members must collectively maximize benefits in resource-scarce environments.[110] Such cohesion played out effectively not only in Ixtlán de Juárez's forestry enterprises but in Rufisque's integration of sanitation and wastewater and in the membership-based organizations established through social movements in Belo Horizonte.

Both community-based and community-driven infrastructures are quite distinct from conventional service-provision models. They turn users into agents, fundamentally transforming the traditional relationship between providers and beneficiaries. It is not mere access to infrastructure services but the transformative power of exercising agency that lights the path to pro-poor economic growth. In fact, the cases in this chapter suggest that synergistic connections between the energy, water, agriculture, transport, and waste sectors often create the conditions under which new microenterprises can take root. When users coproduce infrastructure, they become the agents of self-generated change.

7

Implementing Infrastructural Ecologies: Improving the Odds

The U.N. has put a price tag of accomplishing the new sustainable development goals into the trillions. The first question the world will ask is: "How do you expect to pay for them?"
—Muhammad H. Yunus and Judith Rodin, *Save the World, Turn a Profit*[1]

If the principles of infrastructural ecology are so compelling, why are they deployed so rarely? Why are they the exception rather than the norm? The fact is that in developed and developing nations alike, implementing infrastructural ecology takes time, effort, deep technical knowledge (and its creative application) and, sometimes, more money. But developing countries face particular challenges. First, they are often politically unstable and thus subject to dramatic turnovers in leadership. When new leaders take over, they typically want to be seen as delivering *their* projects, not pushing through the initiatives of the last administration. Second, broadly speaking, the legal, regulatory, and policy environments of emerging economies are less stable, more vulnerable to corruption, and less favorable to innovation. In the face of political and regulatory uncertainty, attempting to attract private companies to help deliver integrated, technically complex projects can be next to impossible.

Given the diversity of actors in the infrastructure space—local and national governments; multilateral banks and other financial institutions; local, national, and international nonprofits (including philanthropies); mission-driven investment funds; private enterprises; and communities themselves—who is best placed to implement new models based on the principles of infrastructural ecology? Government, which has traditionally played a lead role in infrastructure, is often already overburdened, and may be further disadvantaged by a lack of technical expertise. Should a raft of multilateral banks, NGOs, consultants, and global businesses fly in and fly out to deliver the requisite investment?

Should communities simply take matters into their own hands? The answer is yes to all of the above. Although the resulting collaboration may be messy, it becomes more efficient with practice—and eventually proves its value in the form of built projects.

As chapters 2 through 6 demonstrate, an understanding of the conceptual principles of infrastructural ecology allows projects to be conceived, organized, and designed. But good designs are only the first step: the goal is to actually construct the kinds of integrated infrastructural networks that will support sustainable growth. This chapter identifies the obstacles that are typically encountered on the way to realizing built projects and provides suggestions on how they can be overcome. Those who do most of their work in developed countries will recognize many of the obstacles described here—but they are unlikely to have encountered them at the depth that occurs in developing countries, where their apparent intractability often prevents good projects from finding their way to reality.

"How Are We Going to Pay for That?"

If figuring out how to pay for innovative infrastructure projects in developed countries can seem frustrating, attempting to do so in emerging economies can completely confound even the most ambitious infrastructure planners. But as the dozens of case studies throughout this book show, sustainable economic models have proven the financial viability of the infrastructural ecology approach.

The harsh reality is that holistic, sustainable infrastructure projects *are* often more expensive . . . up front. Designing the kinds of integrated systems described in chapter 2, for example, takes longer and requires technologies and expertise that may be significantly more expensive—again, in the short run. Synchronizing approvals and aligning budgets from multiple government agencies requires serious effort (including lobbying)—well beyond what would be required for a single-sector project. Even in the case of the soft-path systems described in chapter 3, construction costs may be higher than those associated with hard infrastructure solutions. And there is no question that business models for energy infrastructure favor electrical generation and vehicle fuels that are carbon-intensive and based on fossil fuels. Finally, the types of meaningful community engagement explored in chapter 6 require additional time and money.

But when project costs are examined over the course of, say, a 50-year life cycle, most of the cost differential evaporates. Moreover, when the value of risk mitigation of the sort described in chapter 5 is taken into account, the balance may be tipped even more strongly toward infrastructural ecologies. Rigorous, holistic cost-benefit analysis that weighs initial extra costs against decades of additional benefits can make infrastructural ecologies look like a sure bet. The deeper one looks into the economics of infrastructural ecology, the more compelling the model becomes.

Combining Multiple Funding Sources

Funding—which is not to be confused with financing (a loan for upfront capital investment, repaid through annual revenues)—is the underlying source of money for initial project costs and operational expenses. Funding for infrastructure projects typically comes from user fees (e.g., electric, water, or trash-collection bills), direct public sector investment (allocations from governmental capital budgets for construction costs), or both. In countries with low per capita incomes (and even lower tax participation), however, funding in any form can be difficult to come by; in these cases, the goal is to use the principles of infrastructural ecology to improve project economics. Multi-sector projects, for example, create potential for multiple revenue streams. Moreover, efficiencies help drive expenses down, and the use of waste as a resource helps increase revenues. Finally, delivering a single project instead of several separate projects reduces overhead and transaction costs.

Improved project economics will attract new types of investors, including private businesses; private global banks; local, national, and international nonprofits; and mission-driven investment funds seeking infrastructure investments that provide long-term returns. At the same time, existing investors, such as multilateral banks, will see the opportunity to make better use of their donor funds by delivering more projects with better outcomes—and will in turn invest even more.

One example of a multisector project is the ecoindustrial park in China's Tianjin Economic-Technological Development Area (TEDA), introduced in chapter 2. At TEDA, treated wastewater is used to cool a cogeneration power plant. Selling treated wastewater provides a new revenue stream for the wastewater treatment plant, but since TEDA charges less for treated wastewater than for potable water, the expenses of the cogeneration plant are reduced, too.

The Clean Development Mechanism (CDM) is another revenue source that has the potential to support infrastructural ecologies. Under the CDM, which was established under the original Kyoto Protocol of 1997, developed countries that have made emission-reduction commitments can implement emission-reduction projects in developing countries. Such projects can earn saleable certified emission reduction (CER) credits that can be counted toward meeting Kyoto targets.[2] The scheme garnered some success through projects in Brazil, China, and India, which earned as much as $20/metric ton of CO_{2e} during the peak years from 2010 to 2011. Unfortunately, European countries were essentially the only buyers for CERs; as a consequence of the European debt crisis, the market for CERs collapsed in 2012 and has not recovered.[3] Attempts to revive the vitality (and price) of the CER market are currently under way. Among the approaches are streamlining documentation and trading processes, allowing fast tracking of common projects (such as microhydro), and simplifying project validation and verification.[4]

No one revenue source will make a project workable, but combining conventional revenue streams with new revenue sources, across multiple sectors, can propel a project into the realm of financial feasibility. Such combined revenue streams allowed the construction of the SMART project in Kuala Lumpur,[5] the Jamuna Bridge in Bangladesh,[6] and Omnigrid Micropower's solar power plant in India.[7]

An interdepartmental budgeting process that looks for overall project benefits and multiple benefit streams can also enhance funding allocations for multisectoral projects. Finding and coordinating capital budget lines from three or four agencies in order to support a single project increases bureaucratic complexity, but it is one means of raising enough money to get a project off the ground. For example, to comprehensively address chronic flooding, Surat, India, established the Surat Climate Change Trust, an inclusive and interdepartmental budgeting process that sought technical solutions and funding sources through new approaches to stormwater management, land-use planning, transportation planning, and even public education campaigns. Today, Surat has a municipal budget line to invest in climate change adaptation, which supports projects that would not otherwise have sufficient funding for implementation.[8]

Securing Predevelopment Funding for Cross-Sector Projects

Most funding and financing sources are looking for projects that meet the following criteria: vetting and permitting are complete, rates of return correspond to total project risk, and market demand is clear and strong. Few funding or financing sources are interested in the predevelopment expenses needed to get from a concept to a fully structured and investment-ready project. Given the complexities associated with infrastructural ecologies, the design and due diligence stages are even more expensive and time consuming—and, as a result, may ultimately be less likely to move forward to implementation. But when such projects do get implemented, the benefits are far greater than those of conventional infrastructure projects.

When it comes to predevelopment funding, infrastructural ecology projects need sponsors who think like venture capital firms. Such firms know that most of the companies they invest in will fail but that a small number will dramatically succeed. The ideal sponsors for infrastructural ecology projects are looking for new solutions and bigger returns on their investments. Unlike private banks, multilateral banks, and the public sector, all of which have limited appetites for risk and failure, nonprofits and philanthropic organizations tend to have the requisite tolerance for higher risk. They are also likely to be seeking new solutions to big problems that often align with the principles of infrastructural ecology. Take the Rockefeller Foundation's 100 Resilient Cities program, for example, which is providing 100 municipal governments around the world with $100 million to research organizational changes, new types of infrastructure projects, and new policies that will help them build more resilient communities.[9]

Multilateral banks, such as the new Asian Infrastructure Investment Bank (AIIB), are particularly well positioned to take a broad view of infrastructure systems and will naturally prefer projects that provide multiple benefits over projects where benefits are few. Because such entities tend to be risk averse—with more interest in not failing than in dramatically succeeding—it may be challenging to obtain funding from them for predevelopment research for infrastructural ecology projects. Nevertheless, if infrastructural ecology projects can be designed and structured effectively, multilateral banks could provide a new source of high-quality projects that will ultimately improve the impact of their investments—both financially and by building long-term resilience. The Urban Climate Change Resilience Trust Fund established by the Asian Development Bank (ADB) is specifically designed to provide predevelopment funding so that projects using the ADB's conventional infrastructure loan programs can find new ways to adapt to climate change.[10]

Finding New Financing Opportunities

In the developed world, once a "bankable" infrastructure project emerges from the feasibility stage, conventional financing sources (e.g., private banks or investment funds) provide loans, and implementation begins. In developing countries, however, infrastructure projects face additional challenges. In particular, to compensate investors for greater exposure to political, social, and financial risk (including currency value, market demand, and construction costs), the rate of return needs to be quite high.

To encourage private sector financing of infrastructure investment, multilateral development banks have traditionally offered credit enhancements—which essentially assign to the multilaterals some of the risks that are specific to developing countries. But some types of risk—such as climate change or the possibility that a stakeholder organization will block a project—simply cannot be mitigated by credit enhancement. They can, however, be mitigated by the inherent characteristics of the infrastructural ecology approach, which, for example, stresses climate-adaptive design and calls for full and meaningful stakeholder engagement. Thus, by their very nature, the principles of infrastructural ecology help reduce some types of project risks and thereby make projects more attractive for conventional financing.

Other objectives of infrastructural ecology—such as carbon reduction—can attract special financing sources that are more mission driven than profit driven. The Kinangop Wind Farm in Kenya, for example—a 60 MW, $150 million facility—is being funded by Norfund, a Norwegian clean-tech fund with a mission to invest $500 million in low-carbon projects in developing countries.[11] Norfund's investment will be paid back over 20 years through the sale of electricity to Kenya Power and Light, the country's electrical distribution utility. Thinking across sectors while keeping climate change mitigation and adaptation (yet another principle of infrastructural ecology) in mind can accelerate outlays for infrastructure in the very places where investment is most needed.

Applying Public-Private Partnership Models

Private sector technology companies such as GE, Mitsubishi, and Siemens sell product lines across multiple infrastructure sectors, including energy, water, transportation, and waste management. And they all have product lines and technologies that are designed to help customers adapt to climate change. So one way to harness new technologies and business-process efficiencies while making optimal use of capital is to partner with technology businesses to design, build, own, and operate infrastructure. Manila Water, which has been awarded the concession to provide drinking water and sanitary water treatment to over six million people in metro Manila, has partnered with Mitsubishi, a Japanese conglomerate with business lines in energy, water, wastewater, and transportation, among other areas. In addition to providing the technological expertise to find efficiencies and synergies between water and wastewater networks, Mitsubishi brings sufficient private sector capital to finance annual expansions of Manila's infrastructure networks.[12]

In another trend sweeping across developing countries, new public-private partnership (PPP) laws are not only allowing but also encouraging private sector investment in public infrastructure. Since national governments are responsible for most infrastructure sectors, they tend to take a broad view, and they are well positioned to create cross-cutting projects that implement the principles of infrastructural ecology. Moreover, some unique public agencies sit atop line agencies—for example, within finance or treasury departments—and are not constrained by single-sector silos and/or geographic territories. National-level PPP laws, when administered by such agencies, should be geared toward projects that serve multiple infrastructure needs—and yield multiple benefits—while requiring minimal public investment. Egypt's new PPP Central Unit, for example, which is housed within the Ministry of Finance, is charged with mobilizing private sector technology, capital, and project delivery expertise, with the goal of producing investment across all aspects of government operations.[13]

Engaging the private sector can be a delicate dance, however, with many potential disadvantages that must be thoroughly considered. Manila Water, for example, has been criticized for having failed to substantially improve water and sanitary systems in its service territories while still enabling private sector shareholders to gain profits. Others contend that Manila Water has indeed brought cross-sector thinking, business-level efficiency, and a level of investment that would not have been possible with sole reliance on the public sector.

Of the many obstacles to implementing integrated infrastructure projects, figuring out how to pay for them can be the most challenging. However, a number of factors—including general economic growth in developing countries and new developments in global financial flows, PPP laws, and the activities of philanthropies and multilateral banks—offer reasons to be hopeful that money can be found to build and operate infrastructure based on new models. Ultimately, paying for infrastructural ecologies will require the same type of creative thinking and risk taking that is required for the design and engineering of the projects themselves.

"Too Slow and Not Our Scope"

All too often, projects are rushed through the earliest and most important stages of scoping, definition, design, cost-benefit analysis, and planning in order to get to the politically cherished groundbreaking ceremony and the laying of the engraved cornerstone. Stakeholders are left out, opportunities to expand and integrate the project get cut off, and agencies plow forward, heads down.

At other times, when integrated, multisector projects appear to benefit other jurisdictions—some of which may be regarded as political foes—there may be a recoiling, a turning inward. Elected officials and the heads of ministries need to be seen as advancing their own agendas for their own constituents, not those of others. And genuine collaboration (across sectors, agencies, or borders) requires participants to give up some of their decision-making authority, and even funding, for what appears to be the benefit of others—resulting in claims that such endeavors represent "mission creep" or are beyond the agency's scope.

Justifying Long-Term Thinking

It is one thing to suggest that projects should reflect longer-term needs but quite another to transform a 50-year life cycle and maintenance plan into actual design briefs, construction documents, and operations manuals. Why would an elected official who is up for election on a four-to-six-year cycle be concerned about such benefits when they may come to fruition under the tenure of the opposition party, which is gaining in the polls? And why spend more now when this year's budget barely meets minimum legal service obligations? Project developers, meanwhile, must answer to lenders and owners who are looking for financial returns within years, not decades. Finally, utilities often claim, rightfully, that the costs of long-term investments will eventually be passed on to consumers in their energy bills, water bills, and bus fares—and that customers are not willing to pay higher rates for such benefits. Ultimately, however, these are short-term, self-centered justifications that lead to higher costs in the long run.

Instead of worrying about whose reign will enjoy the long-term benefits of integrated, holistic projects, elected officials should be using every bit of salesmanship at their disposal to explain to voters that today's investment decisions, if directed toward infrastructural ecologies, will yield many valuable benefits—environmental, economic, and social—in the years to come. Pointing to the visible gains achieved through investments made in decades past usually makes the point clear. Meanwhile, planners and engineers, who may be inclined to focus on climate change mitigation and adaptation, as well as on the many benefits of cross-sector design, should be sure to include comprehensive, long-term cost-benefit analyses along with technically rigorous plans and beautiful renderings. Finally, project developers should demonstrate to lenders the long-term value creation in infrastructural ecology projects. Their pro forma and investment horizons should extend for decades, and they should look for lenders and owners around the world who actually *prefer* long-term profits. Such funding sources do exist, and they are continuously looking for the next project in which to invest.

Through research and publications (including case studies), as well as capacity-building programs, multilateral banks and nonprofits, including philanthropies, should continue to build a body of knowledge elaborating on the benefits of long-term, holistic thinking. Such actors may also be in a position to take on some of the costly and time-consuming public engagement processes without slowing down the project design schedule. Universities and technical training programs can instill in students the value of systems thinking—which, in practical terms, includes taking account of the broadest possible time frame and physical boundaries for study and for engaging in whole-life costing. Everyone has a role in making the best practices of today the typical practices of tomorrow.

Overcoming Governmental Silos

Global corporations spend millions to try to get their highly educated employees to talk to each other across departments. This certainly is not an issue unique to the governments of developing countries. As any public sector employee can tell you, agency responsibilities tend to be clearly and narrowly defined; hence, interagency collaboration can be extraordinarily challenging. It can be hard to get the attention of elected officials on project details—and even if a moment can be found, officials often lack the background or training that would enable them to see the benefits of infrastructural ecology. They just want the project completed on time and on budget. Thus, responsibility for collaboration often falls to a midlevel energy ministry employee who may not have the time, training, or motivation to reach out to a counterpart in a transportation or water department—and may in fact not even know a single person in those agencies, much less the *right* person to collaborate with on a multisector project. Nevertheless, once multilateral banks and philanthropies realized just how many small yet critical decisions are made throughout the ranks of governmental agencies, they began to work directly with midlevel actors—essentially going "under the head" of political leadership to instigate collaboration and coordination between agencies.

Exemplary cross-agency partnerships have generated some of the best models for implementing infrastructural ecologies—demonstrating the potent collaboration that can occur when leaders at the highest levels see the benefits of holistic thinking. Take, for example, the Jamuna Bridge in Bangladesh. Cabinet officials commissioned a study to assess the viability of combining the functions of the bridge's various users (vehicles, a rail line, and a natural gas pipeline) and ultimately created a separate entity—the Multipurpose Bridge Authority—to enable users to negotiate the technical arrangements.[14] Similarly, in New Delhi, India, the local government initiated construction of a tunnel that combines utilities in organized underground conduits, economically serving both the government and the city's commercial center.[15] In yet another example, the Kuala Lumpur SMART project—the award-winning urban bypass tunnel that seasonally stores stormwater, and an inspired collaboration between the Malaysian Highway Authority and the city's Department of Irrigation and Drainage—reaped savings by combining the capital and operating costs of diverse entities.[16] Finally, India's Ministry of New and Renewable Energy, working with the private entity Sardar Sarovar Narmada Nigam Ltd, which developed, owns, and operates an irrigation canal network, created a reciprocally beneficial cross-sector colocation, solar farms set atop canals, which save water while producing electricity.[17]

Crossing Borders

Much has been said throughout this book, particularly in chapters 2 and 3, about the importance of aligning a project's physical boundaries with nature's principles and adopting a technically comprehensive viewpoint. Floodwaters flow downhill without concern for jurisdiction. But what happens when neighboring political leaders are from opposing political parties? Or when development regulations change as one crosses from the city into the county?

Nairobi's efforts to address flooding from a watershed-management perspective high-light the issue of borders. Historically, a stifling blend of city, county, and state agencies had controlled various aspects of flood management within and around the city. In the wake of the devastating floods of 2015, the agencies soon realized that the only solu-tion was to collectively contribute resources and give up some of the decision-making authority within their own jurisdictions. In fact, jurisdictions outside the city—areas that did not have significant flooding problems—would be required to invest within their borders in order to solve flooding problems downstream. And to access state-level funding, downstream jurisdictions would have to allow state-level actors to implement projects within their territory. One can only imagine the political horse trading that must have been required to gain consensus among the assembled participants.

The city-county relationship is often a strained one. The availability of low-cost land drives development outward, causing the counties surrounding cities to bear the brunt of urban expansion. Since the jurisdictions of city agencies end at the city limits, there is little such agencies can do to address the needs of the sprawling informal settlements that spring up on the urban outskirts outside of their jurisdiction. In fact, resources are sometimes erroneously channeled into city agencies to address sprawl when there is little that city agencies can do to influence what goes on within surrounding counties.

Whereas city agencies tend to be better established and have (relatively) more resources—as well as access to initiatives such as 100 Resilient Cities or C40—the pur-view of counties tends to be limited to land issues associated with rural towns or agri-cultural land. Compared with the revenues derived from land sales and property taxes that counties typically see, even the lowest-density, least transit-oriented developments are likely to be viewed as a boon to county tax rolls. As a result, instead of collaborating, cities and their neighboring counties often compete for development and state-issued infrastructure funding.

Similar issues can arise in relation to national boundaries. The most extreme exam-ples are of course carbon emissions and climate change, but other, more localized examples include rivers, lakes, or groundwater that cross national borders. The irony is that in the long run, with well-coordinated planning, a collaborative approach would more efficiently deploy limited resources, ultimately resulting in more and better projects for all parties.

Paying the Price for Neglecting Community Engagement

The fifth objective of infrastructural ecology—meaningful community engagement—has the potential to catch even the biggest and best infrastructure projects off guard. Despite the risks of stakeholder backlash, skipping this crucial step is all too common.

Granted, community reactions can be unpredictable, and meaningful community engagement can slow down design—and ultimately increase the time required to obtain project funding. Moreover, in addition to requiring internal staff resources and funding, the process may require outside specialists who have the training, connections, and experience to conduct community engagement processes. Given limited resources and urgent needs for basic infrastructural services, it may seem to make sense to push toward implementation as quickly and cheaply as possible. But consider the risks of leaving local stakeholders out of the process.

Though the scenario is so common that it barely needs a case study to demonstrate it, take Kenya's Kinangop Wind Farm as one example. For nearly ten years, project developers, including General Electric, engaged in a long and arduous process to create a low-carbon, climate-resilient project—carrying out numerous technical and economic studies and, as described earlier in this chapter, even securing some innovative financing. All approvals, financing, and contracts were in place when the local farming community, on whose land the wind turbines were to be built, filed lawsuits raising concerns about being forced to sell their land and alleging that the wind turbines could cause health problems.[18] After all the hard work, the project now risks being stalled for years. What should a risk-mitigation assessment have looked like during the early stages of that project? What amount of time and what cost would have been required to ensure that stakeholder engagement coincided with design and contract negotiations? And, in retrospect, what would have been the result of a cost-benefit analysis comparing the time and money spent effectively engaging the local community versus the cost of the project's delay or cancellation?

Community engagement is a prime opportunity for nonprofits, philanthropies, and multilateral banks to step in with technical and financial support. Governments and the private sector tend to run on very tight budgets, looking for any opportunity to minimize scope and costs. In contrast, foundations such as the Clinton Global Initiative, the Rockefeller Foundation, and the Gates Foundation, and multilateral banks such as the African Development Bank, the Asian Development Bank, the Inter-American Development Bank, the World Bank, and the new Asian Infrastructure Investment Bank, look for ways to expand a project's scope and to invest donor money to make good projects even better. Stakeholder engagement could be funded through grants or trust funds. Applicants who are aware of the importance of community participation but are unable to persuade the holders of the purse strings to support a project could apply for technical, financial, and capacity-building assistance to ensure integration of this critical component of project design.

In the long run, collaborating with communities does not slow down projects or cost more money; instead, it makes projects better, reduces the risk of delays, and ensures a just process with equitable distribution of benefits. Furthermore, when local communities are informed and involved, they tend to support low-carbon projects that improve environmental conditions, mitigate the effects of climate change, reduce service costs, and provide jobs. But to understand how infrastructural ecology projects can provide such benefits, stakeholders must be meaningfully engaged from the outset. The Jamuna Bridge, described in chapter 2, is an example of a project that follows many of the principles of infrastructural ecologies; however, the community was not fully or meaningfully engaged, and the project has been widely criticized for displacing local residents without due process.[19]

"That's Not How We Do It Here"

How should one respond to the argument that an innovative project based on infrastructural ecology cannot be undertaken because "that's not how we do things here"? Is it the technical education at universities that holds back not only creative thinking but also knowledge of first principles of science and engineering? Or is it the training offered by professional organizations that cling to standard practices, driving solutions toward the patterns established by past projects? Do colonial-era building codes actively interfere with new solutions (as they often do even in developed countries)? Do fuel subsidies and state-owned utilities prevent what would otherwise be economically rational energy- and water-efficient projects from being feasible? Using the principles of infrastructural ecology along with basic scientific and economic principles, we can begin to break down the many barriers to new infrastructural models.

Building Technical Fundamentals

Too often, complex systems founder because local operators lack the capacity to properly operate and maintain them. For example, a technologically unprepared professional class may struggle with modern systems planning and design and fail to adequately train those who are to be involved in ongoing maintenance. For projects to succeed, capacity building must therefore be extended to those at the lowest as well as the highest rungs of the services ladder. One powerful approach to addressing this capacity deficit is through an invested, engaged, and motivated local constituency. Despite having often been marginalized, customers can play a vital role in promoting, developing, and sustaining project integrity.

To optimize energy, water, and waste flows, designers must have a deep understanding of fundamental scientific and engineering principles in the fields of ecology, chemistry, electrical engineering, and hydrology, among others. Alternatively, they need to be able to communicate and collaborate with practitioners in these and related fields. A civil engineer with little or no knowledge of energy systems is unlikely to design an integrated, ecoagricultural project that fully utilizes energy and material flows in the ways described throughout this book. But a civil engineer who is inspired by multiple disciplines and actively searches for cross-sector opportunities might propose an irrigation canal shaded by solar panels that provide rural electrification.[20]

Universities certainly have a big role to play in building capacity in technical fundamentals, as well as in the "softer" aspects of infrastructure, such as the value of integrating multisector synergies, community engagement, and long-term economic development into design. Urbanizing settlements need graduates who think creatively and understand social, economic, political, environmental, and infrastructural interrelationships so that they can develop new solutions that reflect and respond to the local conditions in which they operate.

As young people progress through their careers, encountering new ideas and new models for infrastructure development, they need to exercise influence through professional organizations. But professional bodies—such as architectural, engineering, or urban-planning accreditation entities—tend to perpetuate single-discipline training based on past solutions instead of demanding breadth of knowledge and creative solutions. At the same time, professional organizations are well placed to ensure that infrastructural ecologies are adapted to local conditions—and that practitioners in a given region fully understand and implement global best practices. Professional bodies often run continuing-education programs for leading firms, agencies, and organizations that make critical infrastructural decisions, as well as for designers and policy makers who are well along in their careers; such training programs are well positioned to advance new models of infrastructure development.

For infrastructural ecologies to become the norm rather than the exception, there is also a need for cross-cutting organizations that bring together a variety of professions, viewpoints, and stakeholders to share ideas outside the highly charged environments created by actual projects or proposed legislation. The Urban Land Institute (ULI), based in Washington, D.C., is an excellent example of such an organization, as its mission is to engage the members of many different professions—including property developers, public sector employees, designers, lawyers, and engineers—in an open, public dialogue on the responsible use of land. ULI has successfully expanded to Europe and Asia and is now looking for more opportunities around the world to advocate for mixed-use, transit-oriented urban development. ICLEI, C40, UN Habitat, the Organisation for Economic Cooperation and Development (OECD), and the World Bank Group are also excellent examples of cross-cutting organizations that practice globally and can support emerging economies in their quest for improved technical training for professionals.

Rebuilding Building Codes

The building code in Sri Lanka was last updated in 1978 and is based on technologies and best practices that date from the 1960s. In Lebanon, a water-scarce country, none of the laws, standards, or regulations for water management even consider the possibility of wastewater reuse.[21] Is it not time to update building codes across the globe in order to foster new, sustainable development models that have been proven to lower life-cycle costs while making buildings healthier and safer?

Storms are intensifying around the world, so drainage designs need to be revamped to cope with heavier water flows. One-hundred-year flood levels are rising, so ground-floor elevations and electromechanical systems need to rise accordingly. Peak daytime temperatures are also increasing, so everything from air-conditioning systems, to public parks, to hospitals needs to be capable of responding to changing conditions.

Many other climatic, economic, technological, and social parameters are different today than they were 20 years ago. And in 20 years, they will be different from what they are now. Thus, it is not enough to address today's conditions. As we build infrastructure, buildings, and settlements that will last for 50, 100, 200 or more years, we need to consider tomorrow's conditions. In the same way that fire and life-safety codes, accessibility requirements, and earthquake design standards have been updated and integrated into minimum requirements based on global best practices, sustainable design and resilience planning should be incorporated into minimum standards.

The particular issues that need to be addressed vary with location, but all municipalities have one thing in common: they need a regular process for updating building codes and infrastructure design standards at least every five to ten years. As part of their review of codes and standards, state and municipal governments should undertake a broad stakeholder engagement process, calling for recommendations from developers, construction companies, members of the design profession, community residents, and even global professional institutions such as the Royal Institute of British Architects

(RIBA), the International Standards Organization (ISO), the International Building Code (IBC), and the American Society of Civil Engineers (ASCE). There are even nonprofit institutions, such as the International Partnership for Energy Efficiency Cooperation, whose sole purpose is to advocate for better building codes around the world.

The next stage is to follow through with educational outreach to builders and design professionals to ensure that everyday design practices reflect the updates. Ideally, building codes should encourage or mandate the principles of infrastructural ecology, but at the very least, they should not interfere with their implementation.

"Will the Next Administration Support This?"

It is no wonder that infrastructure projects get pushed through the design and approval processes as quickly as possible in order to get that cornerstone laid before the next election. When projects take years or even decades to implement, it can be challenging to maintain continuity across administrations. A mayor or governor who ran a successful campaign by calling for change, for example, may question funding for a long-term infrastructure project. Continuity may also be affected by shifts in the legal or regulatory environment, or even by a subsequent administration's corruption.

Creating a Predictable Regulatory Environment

Most utility companies must appease a complex mix of entities: profit-seeking shareholders, governmental agencies, commercial and industrial customers, and residential customers. Above these networks are regulators, whose role is to ensure that all actors play fairly and abide by local, state, and national regulations and policies. Investment horizons, capital expenditures, operations and maintenance budgets, and utility rates are subject to strict review and must be approved by regulatory boards. Thus, for water infrastructure to be shared with, say, transportation infrastructure, the arrangement must be approved by both water and transportation regulators and then integrated into the cost of water and the cost of bus or train fares in a way that is fair to utilities and consumers. This is a complicated process indeed, but it *has* been done—and should be done more often.

Public and private utilities alike often accuse regulators of stifling innovation—including holistic, long-term thinking. Nevertheless, regulatory bodies may be particularly well positioned to lead the charge to advocate for, and even require, projects that are based on the principles of infrastructural ecology. For example, water utility regulators may begin to require the operators of wastewater treatment plants to collaborate with drinking water utilities in order to safely, beneficially, and cost-effectively integrate water reuse into their systems.

If regulators do begin to propose new models for infrastructure, it must be through an open, transparent, and publicized process that invites and considers the perspectives of all affected parties. Take Singapore's Public Utilities Board (PUB), for instance—a single entity that operates and manages three highly interdependent water sectors: drinking water, stormwater, and wastewater. Thanks to effective regulatory authority and technical capacity, as well as strong political backing, the PUB has been able to address issues as complex and varied as water security, demand management, affordability, and equity of access.[22]

When elected officials call for new models of infrastructure delivery, regulatory bodies may be placed in a difficult position. On the one hand, they need to respond to the request; on the other, they must continue to demonstrate the stability of the utility system and avoid being seen as entirely subject to the whims of each new administration. It is possible to find middle ground, however, and to balance the need for reform with the need for stability. Reform processes that follow deliberate, transparent, and inclusive procedures will allow the principles of infrastructural ecology to enliven the work of regulatory bodies around the world.

Gauging Legal Frameworks

Aside from the several examples of grassroots, self-built infrastructure projects included in this book, nearly all infrastructure requires some investment, risk, and contractual obligations. And for projects that require financing of any sort, basic legal frameworks must be in place. Property laws must be sufficiently strong and well defined that legal claims on the property do not pop up during construction (or even further into the future). Contract law must have a sufficient track record in upholding agreed terms in cases where disputes have arisen. Tax laws must be clear and stable enough to allow project budgets to incorporate all domestic taxes and consistently follow international tax treaties. For public projects, procurement laws must provide transparency and allow sufficient flexibility to incorporate new technologies and new solutions. Considering the added scope, complexity, and time associated with infrastructural ecologies, secure legal frameworks become all the more important.

The stability and enforceability of legal frameworks in any given country or region may guide project scale, structure, and complexity. Where legal structures are weak (as in Burundi, Haiti, and Nigeria, to name a few examples from this book), project developers may be more inclined to advocate self-built and community-owned infrastructural ecologies. Developing countries with stronger legal structures (Morocco, Mexico, and South Africa, for example), in contrast, may lend themselves to more complex financial and ownership models that draw on international institutions and companies.

In sum, long-term projects with global participants require confidence that contracts will be upheld for decades, that tax laws will remain fairly stable, and that property laws will be maintained. The World Bank's annual publication *Doing Business* summarizes and ranks all countries on a variety of factors relating to legal frameworks.

Sidelining Corruption

Corruption is well known to hamper economic development, restrict foreign direct investment, and disenfranchise communities.[23] It does not—as some proponents suggest—grease the wheels of the free market.[24] And, like other obstacles to implementation discussed in this chapter, it is not a scourge only in developing countries; it can be found around the world, at all levels of business and government.

Why does corruption interfere with the implementation of infrastructural ecologies? The U.S. Foreign Corrupt Practices Act (FCPA) criminalizes bribery of foreign officials in exchange for business opportunities, levying major fines on companies or individuals found to be engaging in such practices.[25] The FCPA and other international corruption laws are rigorously enforced—and they place the responsibility for avoiding corruption on those who are seeking to do business rather than on the officials seeking bribes.[26] As a result, where global technology companies, for example, might otherwise consider partnering with the governments of developing countries, the (arguably well earned) perception that corruption is part of doing business in such locales prevents such companies from even entertaining the idea of investing. Meanwhile, private global banks have become so laden with regulations developed to track the flow of funds for terrorism, corruption, and economic misdealing that very few can do business in new or emerging markets. So when the call goes out for new models of infrastructure that require technical capacity, specialized technologies, capital financing, and global collaboration from businesses and individuals based in developed countries, it is no wonder that progress is slow.

Multilateral organizations already play a big role in assisting governments in overcoming the barrier of real or perceived corruption by facilitating and endorsing transactions while adhering to the strictest principles of transparency and honest dealing. To ensure the legitimacy of the development process, the governments of emerging economies often approach institutions such as the World Bank and the International Finance Corporation, the World Bank's private sector arm, to procure projects. One of the obstacles to full U.S. support for China's new Asian Infrastructure Investment Bank (AIIB) was the lack of procurement transparency and the absence of procedures to ensure honest dealing. By rectifying these problems, the AIIB could demonstrate best practices not only with regard to new models for infrastructure design and financing but also in procurement and implementation.

When corruption and misdealing influence infrastructure planning, communities are the real losers; they are also the stakeholders who have the most to gain by speaking up and demanding a clean process. Holding officials accountable can be challenging, but grassroots efforts can be successful. There are plenty of strategies available for tamping down corruption, but the first step is awareness, followed by real effort by all parties to move toward transparency and legitimacy.

Ways Forward

By increasing investment, access to basic services, and equitable distribution of benefits, infrastructural ecologies have the potential to help address the political and economic instability that plagues many developing countries. Looking across the globe at the approximately 145 countries that the United Nations categorizes as "developing," there are plenty of bright spots where global businesses and local communities do feel safe making substantial investments for the long term. These should be the target areas for implementing infrastructural ecologies in the short term.

Putting the Five Objectives into Practice

Ecological design . . . is the careful meshing of human purposes with the larger patterns and flows of the natural world and the study of those patterns and flows to inform human actions.
—David W. Orr, *The Nature of Design*

Of all the complex development challenges faced by the countries of the Global South, the infrastructure gap is the most pressing. Despite increasing economic competitiveness, emerging economies lack the critical services needed to ensure the welfare of their citizenry. No one book can delineate the myriad strategies that will be needed to close this gap. What infrastructural ecology offers, however, is a place to start—a novel platform that can generate efficient, resilient solutions with less environmental impact and a more equitable distribution of benefits. But because infrastructural ecology is somewhat at odds with conventional, sector-based approaches to public works, implementation requires a combination of transformational leadership and innovative policies. Focusing on each objective of infrastructural ecology in turn, this chapter offers specific practical recommendations to support the transition to an infrastructural ecology approach.

Objective 1: Relational Solutions

Countering the historical fragmentation of infrastructure development requires a collaborative, holistic approach that reaches across utility sectors; includes all levels of government; and embraces private actors, the development community, and community stakeholders. In theory, integration can start anywhere. The Programme for Infrastructure Development in Africa (PIDA), for example, is a market-driven initiative that promotes fiber-optic investment. Under this program, fiber-optic lines share underground conduits and overhead line space with the power grid, thereby taking advantage of existing power lines as well as new road or rail right-of-way construction.[1] In one case, coordinating network implementation with road construction added no more than 2 percent to the total cost of the road while reducing the network operator's costs by 80 percent.[2]

The low marginal costs of adding spare ducts, pipes, or cable capacity to road or rail construction offers the potential to share the savings with the power or transport sector—or, as in the case of PIDA, to reinvest by extending service to remote rural areas, where expansion would otherwise be more costly. Such "disruptive," market-driven efforts are laying the groundwork for comparable infrastructure-sharing strategies and policies worldwide;[3] for example, transmission towers are already being shared in Brazil, India, and Indonesia.[4]

The cases in this book demonstrate that many of the most robust infrastructural ecology projects have been driven by opportunistic thinking, whether on the part of private sector entities, civil society, or other actors. Nevertheless, government must often take the lead. In "Systems Thinking and the Future of Cities," author and activist David Orr argues that managing "the complex cause and effect relationships between social and ecological phenomena" requires a systems approach to urban governance— "new habits of mind that see interrelationships, system structure and patterns."[5] Governmental authorities may be skeptical of disruptive innovations, but they are also likely to be interested in the results—and may therefore be willing to acquire the new habits that are needed to support infrastructural ecology.

The next two sections offer recommendations for governmental actors who wish to engage in an integrated, holistic approach to planning, along with procedures that can help lead actors (in any sector) to implement integration in specific projects.

Fostering Integrated Planning and Implementation

As chapter 7 notes, governments (at all levels) are often in the best position to coordinate the many entities that are typically involved in infrastructural implementation. The following are useful strategies for aligning the objectives and activities of ministries and departments across geographic and jurisdictional boundaries.

• **Engage staff from different ministries and/or departments in new approaches to cross-sector infrastructural development.**
   • To break through entrenched patterns, ask staff to jointly develop cross-sector objectives that can be implemented through partnerships on pilot projects.

• Evaluate multiple projects for potential complementarity and synergies, as well as for adverse impacts, selecting for implementation those that have the potential to be replicated regionally or nationally.

• Create a separate authority or commission that is empowered to lead integrated resource planning for interdependent sectors, such as energy, water, and agriculture.

• Require that all cost-benefit analyses incorporate the economic value of natural resources.

• To ease negotiations or resolve disputes, consider using outside facilitators.

• To encourage a systems-thinking approach, foster discussion among multidisciplinary teams of planners, engineers, and design professionals.

• **Invest in coordinated mapping and databases.** To reveal potential synergies, create interdepartmental access to geospatial information on transportation and utility grids, centralized assets, and distribution networks.

• **Create economic and regulatory incentives.** To encourage projects modeled on infrastructural ecology to be funded by local government, link implementation with access to national and/or international development funds. To reward cross-sector planning and implementation on the part of private or nongovernmental entities, governments should use instruments such as expedited approvals or tax breaks.

Developing and Implementing Infrastructural Ecology Projects

Project-level development and implementation requires team building, agreements, and analysis of potential synergies. The following recommendations will help ensure successful cross-sector collaborations during the project implementation phase.[6]

• **Engage the right actors.** Ensure that the team overseeing implementation includes members from different sectors, different levels of government, and civil society. Obtain buy-in from executive-level decision makers (whether government or corporate) as early in the process as possible.

• **Look for opportunities for synergy.** Using data on each system's resource flows, identify potential opportunities for synergies—such as energy, nutrient, or water cascades.

• **Use comprehensive, long-term cost-benefit analysis to evaluate complementary services, products, facilities, and assets.** Where possible, assign a dollar value to benefits such as increased productivity or resilience.

• **Engage multidisciplinary design and engineering firms.** To encourage innovation, consider using a design competition or a performance-based design/build contract. Under the design/build approach, the provider is required to meet specified performance goals but is free to determine the methods used to do so.

• **Use consensus-based decision making.** Outside facilitators can help ensure the efficiency of consensus-based processes.
• **Lock in commitment to identified goals.** Develop formal agreements for institutional collaboration that reflect each party's needs and capacities.
• **Monitor performance and document outcomes.**
• **Train all those who engage with the new systems in new modes of use and shared responsibilities.** Training—for operations and maintenance staff, for example, as well as for users—is key to obtaining the desired performance, savings, and benefits.

Objective 2: Ecological Alignments

Infrastructural ecologies can reduce direct, indirect, and cumulative impacts on natural capital. Low-impact, soft-path alternatives are especially relevant to the integrity and resilience of water services. They can pave the way for both natural and constructed systems that are culturally appropriate and climate-adaptive, and they can also regenerate ecosystem services and thereby improve the level of well-being per unit of water used.[7] Finally, addressing the interdependence of water services—for example, the procurement and distribution of freshwater, and the management and recycling of storm- and wastewater—can help speed the transition to integrated water systems.

Thanks to the Integrated Water Resources Management (IWRM) framework, which was designed by the Global Water Partnership to replace traditional fragmented solutions, the water sector worldwide has already begun to apply an integrated, soft-path approach.[8] Under the IWRM, planners and managers are looking at the cycle of supply, treatment, and reuse more holistically, at watershed scales. They are also studying the efficiencies that can be obtained through multiple uses. These include irrigation for food security, industrial use, and essential support of natural systems. Finally, alternative modes and scales of operation—beyond traditional centralized services—are being used more widely. These localized (distributed) systems have several advantages over centralized solutions. First, they collect, process, and recover not only water but also beneficial nutrients. Second, they can be ramped up more readily than conventional facilities, either as stand-alone assets or as components of centralized systems. ("Polycentric" service arrangements, which combine centralized and distributed approaches, are more resilient than centralized systems alone.)

As discussed in chapter 5, there is increasing competition between the energy and agricultural sectors for access to water. Thus, the IWRM framework must address cross-sector interdependence. Despite the potential for conflict, the food, water, and energy nexus also offers opportunities for synergy. For example, waste flows from water treatment can be recovered for energy production or agricultural use.

The recommendations that follow are designed to assist governmental water ministries and departments as well as private sector operators in developing integrated, holistic approaches to water supply, drainage and stormwater management, and treatment for recovery and reuse.

Integrating Capture and Storage

Authorities in all urbanizing areas—not just those in water-stressed or water-scarce regions—need to ensure secure water supplies for burgeoning populations. Public policies, including incentives for private operators, can create alternative supply sources by optimizing both water catchment and reservoir storage capacity.

• **Mandate rainwater harvesting.** Rooftop rainwater harvesting can help alleviate scarcity while reducing runoff. This distributed approach is equally applicable in urban and rural areas characterized by highly variable rainfall.

• **Promote aquifer restoration.** To recharge depleted aquifers, use drainage systems to capture and infiltrate runoff that would otherwise be wasted.

• **Reduce evaporation and promote infiltration.** To reduce loss from evaporation, cover or shade reservoirs. Amplify local infiltration by establishing catchment at a range of scales.

Using Green Infrastructure to Manage Drainage and Capture Water

A wealth of low-tech, green infrastructure strategies can assist with climate adaptation, flood abatement, and stormwater management, while enhancing the urban environment. To take advantage of these options, water ministries and departments need to formulate policies and regulatory structures that will encourage private investment in such infrastructure. They also need to be creative in combining green infrastructure with internally funded or internationally funded conservation-related initiatives and in merging project funds accordingly.

• **Design drainage systems that direct stormwater to catchment areas or infiltration zones.** Develop or revive tank systems like those that have historically been used in India. Establish diffuse, localized infiltration measures (like Indonesia's *biopori* system) that not only alleviate stormwater problems and restore groundwater but also yield agricultural fertilizer.

• **Assign priority to projects that yield multiple benefits.** To optimize cross-sector benefits—such as improved public health and agricultural yields—provide incentives to the civil and private sectors that will encourage multifunctional infrastructure. The enlargement of the Iguaçu River basin in Curitiba, Brazil, for example, doubles as both a flood-control measure and an urban park system, and it was undertaken using federal funds that would otherwise have been dedicated to hard-path systems.

• **Partner with other entities to fund water projects.** Where possible, use monies intended for constructed systems to fund green infrastructure. These can be combined with funds from departments charged with conservation, reforestation, recreation, biodiversity protection, and even adaptation.

Recovering and Reusing Water and Nutrients

Water ministries must move away from the linear, "once-through" systems that encourage waste of valuable water. Instead, biological processes should be used to recycle wastewater for nonpotable uses and nutrient recovery, either for agricultural purposes or for conversion to methane. Transitioning to such approaches will require careful consideration of various options for end use, as well as collaboration across sectors and jurisdictions and between governments, nongovernmental organizations (NGOs), and private operators.

Santiago, Chile, and the kingdom of Lesotho reveal the different scales at which such recovery systems can operate. In Santiago, the wastewater treatment plant recovers methane, bringing it to pipeline quality for municipal energy use. In Lesotho, domestic dome digesters improve sanitation while producing valuable biogas and fertilizer.

• **Integrate horizontally and vertically.** To create an enabling environment for water recovery and reuse, ensure that national policies, legal frameworks, data collection, public education, and research and development are aligned—not only for the water sector but also for related domains such as energy, sanitation, and agriculture.[9] To support cross-sector integration, use incentives—including subsidies, price controls, and pollution taxes. In Nashik, India, for example, the elimination of the energy subsidies that had encouraged excessive pumping for irrigation, along with mandated pump upgrades, conserved both energy and water.

• **Consider recycling wastewater.** Particularly in arid climates (Jordan is one example), treated, nonpotable wastewater can be a reliable, reasonably priced alternative water source for industry, sanitation, or air-conditioning.

• **Consider reuse of wastewater for agriculture.** Reclaim wastewater to maintain or replenish soil fertility, reduce pollution of surface waters, offset the use of synthetic fertilizers, and conserve potable sources for higher use. To eliminate health risks, ensure that projects meet current protective standards.

• **In developing policies and practices, take end users into account.** Convene participatory, cross-sector stakeholder groups to discuss cultural and practical preferences with regulators. Educate consumers to ensure the appropriate use of diverse water sources.

Objective 3: Low-Carbon Processes

Given current carbon-reduction strategies, annual global emissions may (at best) level out at 50 billion metric tons (55.1 billion tons) of CO_2 equivalents in the decades to come, although there is a risk that emissions may go even higher.[10] These projections evoke a not-so-distant future fraught with climate perturbations that cannot be forestalled. As the demand for power and transport in developing nations escalates, the rapid development of new, low-carbon economies is imperative and must be supported by the countries of the Global North.

The good news is that developing nations can be swiftly primed to decouple their economic output from carbon-based energy consumption. This transition would entail the use of renewable resources—solar, wind, microhydro, and waste organic material— along with the development of diverse low-emission fuels for power, heat, and transport. Significantly, transforming the carbon trajectory of emerging economies will call for deliberate action on the part of governments globally, the NGOs, the private sector, and individuals.

On the one hand, costs for renewable power have dropped to the point where, at least in some locations, renewables are competitive with petroleum or even gas.[11] On the other hand, many developing countries lack the technical and administrative capacity to install and maintain renewables. The challenge is to decarbonize while using systems that are bankable, scalable, and provide equitable access.[12]

In urbanizing areas, the goal is to avoid locking in decades-long commitments to coal-fired generation while still producing sufficient power to meet growing industrial, commercial, and residential needs. Moving away from coal and other carbon-intensive energy sources will require generation from wind or solar farms on a power plant scale,[13] as well as transitional measures such as gas-turbine generation of combined heat and power. Nor are wind and solar the only renewables available to emerging economies. Landfill gas, cooking fat, and wastes recovered from humans, animals, and agriculture are being successfully utilized. Fortunately, many emerging economies are located in climate zones where biogas systems (both at household and commercial scales) perform best; moreover, implementation of such systems is largely within local means. Thus, ideal solutions—either permanent or transitional until grid connection is available—are already available.

Despite some encouraging signs, stronger government policies and strategies are needed to effectively decarbonize power, heat, and transportation. The five subsections that follow offer recommendations for achieving carbon-reduction goals.

Integrating the Renewable Energy Sector

Energy markets in emerging economies tend to be shaped by many actors operating at multiple levels—from local entrepreneurs who supply household photovoltaic power, to state-owned utilities supplying their customers with distributed power or grid-connected access, to national and international development agencies financing large, grid-connected systems.[14] Given such decentralized patterns of production and consumption—which are accompanied by relatively weak financial markets and underdeveloped legal and regulatory frameworks—the governments of emerging economies face significant challenges in their efforts to integrate fragmented energy-delivery models while transitioning to lower-carbon technologies. National renewable-energy policies will be crucial to overcoming these difficulties and successfully developing indigenous renewable resources.

Promoting Renewable Generation

Many low-latitude countries have abundant potential for renewable energy aside from hydroelectricity. Governments can take a number of key steps to promote renewable power generation:

• **Retire carbon-based subsidies.** Stopping the importation of "cheap" oil and gas can create incentives for the development of indigenous energy sources while also improving the balance of trade.

• **Obtain support from the developed world.** To foster the development of renewable energy resources, emerging economies need professional education and practical training. In most cases, such capacity building will require outside support.

• **Encourage participation of the private and civil sectors.** Corporate developers and nongovernmental agencies can be key to the development of both small-scale, distributed energy systems and larger-scale, centralized arrangements. But they will need incentives to undertake such projects.[15]

Expanding the Use of Alternative Heating Technologies

Many developing countries are embracing, and even mandating, solar water heating—a simple, low-cost technology. The heat output of China's approximately 30 million rooftop installations is currently equivalent to that of 49 coal-fired power plants.[16] In Brazil, 22 cities have made solar thermal technology compulsory for new construction, using incentive programs to soften the impact; similar legislation relying on mandates and incentive systems is under consideration in 80 other cities.[17]

The heating sector in the Global South is largely undeveloped—partly because of low demand and partly because the millions of installations that do exist are geographically diffuse and thus more difficult to regulate (and less responsive to incentives).[18] Where winter warming is required, however, government policies should support renewable technologies such as biomass and solar thermal, both of which can help reduce reliance on imported fossil fuels. In the case of both heat and hot water, a transparent policy framework allows governments to signal industry and other heat consumers about the opportunities for and benefits of moving away from carbon-intensive fuels.[19]

Recovering Waste for Energy

In the Global South, between 30 percent and 60 percent of municipal solid waste is still openly burned or dumped.[20] Although more and more cities are improving their solid waste disposal practices by establishing sanitary landfills, that approach contributes to yet another problem in that landfills are a major source of methane emissions, which are projected to rise 19 percent by 2020.[21]

There are three compelling reasons for governments to mandate or provide incentives for the long-term, on-site recovery of landfill gas: first, to eliminate emissions; second, to obtain low-cost renewable energy; and third, to qualify for carbon credits under the Clean Development Mechanism, which allows developing nations to finance emission-reduction projects by selling credits to industrialized nations that are attempting to meet their own carbon-abatement targets. To accelerate urban reclamation of municipal solid waste, organic agricultural waste, and sewage gas, governments need to engage in high-level policy integration, create new partnerships, use incentives, educate their citizens, and consider next-generation technologies. The following recommendations address those goals:

• **Establish joint water, energy, and agriculture commissions.** In Nepal, for example, the creation of the Biogas Support Program opened up opportunities for private biogas companies to scale up rapidly.[22]

• **Create new, cross-sector partnerships.** To make beneficial use of treated methane, government agencies can encourage cooperation between the energy sector and public and private sector landfill operators.

• **Support energy recovery through a combination of direct funding and incentives.** In both rural and periurban areas, government agencies can (1) provide incentives for biogas capture from animal, human, and agricultural waste and (2) encourage the establishment of community- or district-scale biodigesters.[23]

• **Use education to support changes in perception and behavior.** Cultural aversion to handling or processing human or animal waste may lead to resistance. The implementation, management, and scaling up of biogas programs will therefore require changes in behavior (ability to collect and transfer waste). Education—both in the schools and through the dissemination of information by public health ministries—is key to fostering the necessary cultural and behavioral shifts.

Adopting Low-Carbon Transport

In emerging economies, curtailing the boom in private vehicle ownership is key to reducing greenhouse gas emissions. To discourage reliance on automobiles, governments need to foster nonmotorized transport while ensuring that public transportation is efficient, affordable, and appealing. The following recommendations address the kinds of government policies that will be required to achieve those goals:

• **Integrate policies across all levels of government.** Fostering public transportation requires commitment at the national, state, and local levels. Three strategies are key: (1) incorporating transit access into land-use planning; (2) developing new intermodal transit facilities; and (3) undertaking improvements of the "last mile."

• **Use education to encourage behavioral change.** Quantify and publicize the benefits of low-carbon transport: mitigation of air pollution, water pollution, and the heat island effect; reduced vehicular congestion; and lower cooling costs. Foster understanding of the global benefits of reducing carbon emissions—in particular, the connection between carbon-intensive transportation and climate change.

• **Promote nonmotorized transport.** To improve safety, add dedicated lanes for nonmotorized transport to new or existing rights-of-way.

Objective 4: Resilient Constructions

Developing countries are among those most at risk from the effects of climate change, and their lack of financial, technical, and administrative capacity renders them even more vulnerable to flooding, storm surges, rising temperatures, and water stress. In nations situated in the tropics or in low-latitude dry zones, rising temperatures are stifling urban areas, damaging road and rail infrastructure, and contributing to public health crises. Meanwhile, freshwater supplies are threatened by accelerating desertification in drought-prone areas of the interior and by the increasing salinization of coastal aquifers. Critical steps for adaptation include developing national policies, fostering implementation of both constructed and natural coastal protections, promoting green infrastructure in the interior, and implementing water security measures.

Creating a Consistent, Nationally Driven Climate Policy

Given the diversity of vulnerabilities and their geographic dispersion, adaptations have rarely been coordinated. But climate change requires an integrated, national-level approach, localized action plans, and the integration of climate-adaptive codes and standards.

• **Develop coordinated national policies.** To counter fragmentation, engage in cross-sector strategic planning and policy alignment, including not only the water, energy, sanitation, land-use, and transportation ministries but also those charged with health, forestry, and agriculture.[24]

• **Mandate or encourage the development of climate action plans.** Subnational government- and stakeholder-driven climate action plans should draw on local knowledge; incorporate cross-sector policy measures and technologies; and assign priority to measures that meet multiple objectives—such as diverting floodwaters to create rainwater storage and incorporating drainage into new road construction.

• **Integrate adaptation measures into national codes and standards.** The least expensive path to climate adaptation is to incorporate resilience from the start; thus, updated codes and standards that mandate resilient features are key.[25]

Combining Hard and Soft Measures for Coastal Protection

National and subnational measures to address flooding, erosion, and submergence should be designed to reduce risk to populations, be geographically integrated, and be resilient.

• **Reduce population vulnerability.** First and foremost, land-use strategies and regulations should be designed to divert populations from flood-prone areas.

• **Develop integrated coastal zone management strategies.** Deltaic, coastal, and interior areas must be addressed in relation to each other. Integrated measures include the following:

• Upstream diversion and/or retention
• Dredging and fortifying canals and rivers to improve the water storage capacity of receiving waters
• Strengthening natural intertidal zones through plantings
• As a last (and most costly) defense, the construction of seawalls
• **Consider floating structures.** Where other measures will not suffice, create buoyant platforms that can help support agriculture while also providing refuge.

Addressing Warming through Adaptation and Mitigation

Land-use planning, urban design, building practices, and civil engineering typically fall within the purview of municipal authorities. But the urban fabric of informal settlements has largely been created by the inhabitants themselves. The resulting building density, proliferation of impervious ground surfaces, elimination of open space and plant cover, and loss of bodies of water exacerbate the urban heat island effect. To support adaptation and mitigation, governments should promote cross-sector strategies, encourage civic participation, and promote the use of waste heat.

• **Link policies that reduce traffic congestion and the urban heat island effect.** Reducing congestion decreases heat-trapping gases and thereby lowers the energy demand associated with air-conditioning. In Curitiba, Brazil, for example, the development of the rapid-transit bus system included extensive tree plantings that now overarch the dedicated rights-of-way. As a result, Curitiba experienced a reduction in the urban heat island effect and enjoys Brazil's lowest rate of urban air pollution.
• **Foster civic participation in green infrastructure development.** Encourage or require businesses and residents to plant trees and other greenery at both the street and building scales. Fast-growing creepers and vines can be readily grown in yards and on buildings, and street plantings have the added benefit of reducing flooding and roadbed erosion.
• **Promote the use of waste heat.** In the case of large commercial or institutional facilities, the heated exhaust from air conditioning can be recovered for domestic hot water heating.

Increasing Water Security

In regions that have already been designated as water stressed or water scarce, further disruptions in the hydrologic cycle are expected. As lower rainfall and expanding desertification jeopardize water security in both urban and rural areas, especially those lacking piped water, governments and other actors must take action.

Increasing water security will require a deeper understanding of interconnected use patterns—in particular, competing demands in the food, water, and energy nexus. Working together, governments and civil society must amplify storage systems, create seasonal surpluses to replenish aquifers, reverse desertification, and evaluate desalination options.

• **Optimize storage of seasonal water.** Seasonal water stored in soil or aquifers or in constructed bodies of water such as cisterns can offset the overabstraction of groundwater that plagues much of the developing world, smoothing out variations in watershed resources.

• **Use reclaimed wastewater to replenish aquifers.** Treated wastewater can be directed to aquifers for crop irrigation, as it is in Jordan, Mexico, and much of Africa—an approach that may become standard practice in periurban regions, where urban wastewater can be exported to nearby agricultural areas. Among the benefits are opportunities for broader replenishment of ecosystem services.

• **Stabilize grasslands against desertification.** In many parts of the world, land-use changes have eroded topsoil and disturbed rainfall patterns. To restore grasslands and improve water retention, consider grazing practices that mimic instinctive herd behavior, which both fertilizes and aerates the soil.

• **In hot, dry regions with access to the sea, consider solar-powered seawater desalination.** Among the options for water production are photovoltaic desalination plants and seawater greenhouses.

Objective 5: Codevelopment

Emerging economies are struggling to extend infrastructure services to their entire populations and to do so equitably. Increasingly, services developed through local participatory planning are yielding better and more sustainable results, particularly in rural areas. Inclusive processes also have added benefits, including poverty alleviation, capacity building, and greater self-reliance. Moreover, the integration of local knowledge, cultural preferences, and indigenous skills can strengthen the sense of ownership of coconstructed systems.

Policies and practices to improve community participation in infrastructure planning and operations include decentralizing decision making, mainstreaming community participation, and supporting entrepreneurship.

Decentralizing Decision Making

Developing countries face severe limits on national capacity, not only in raising capital but also in providing efficient and equitable delivery of public services. As one strategy for addressing these issues, many development agencies are recommending devolving infrastructure planning and implementation to subnational levels. Among the advantages of devolution are increased accountability to customers and the community at large, as well as greater ability to accommodate regional or cultural preferences. Where decision making is decentralized, however, both national and local governments need to ensure that measures are in place to prevent domination by local elites; such measures include meaningful democratic participation and increased transparency.[26]

Where local democracy does not yet function effectively, progress toward decentralization may be postponed. In some areas, including Viet Nam and Latin America, decentralized infrastructure planning and implementation has been relatively successful.

Mainstreaming Community Participation

Directly involving stakeholders in infrastructural decision making is no longer a novel concept. Nevertheless, some observers feel that in many emerging economies, genuine participation continues to be weakly implemented. Governments, the private sector, and NGOs need to more effectively engage local decision makers and advisors; make information accessible; clearly establish the distribution of costs, benefits, and risk; and facilitate behavioral change.

• **Engage local decision makers and advisors.** To protect against domination by any one stakeholder group, provide for democratic election of local participants. And to ensure that design and implementation are compatible with local beliefs and practices, include local lay advisors and planners, as well as analysts, engineers, and NGO staff.
• **Make information accessible.** At each phase of planning, consensus building, negotiation, and agreement, confirm that all stakeholders, including community members, fully understand complex information.
• **Clearly establish the distribution of costs, benefits, and risks.**
• **Facilitate behavioral change.** Successful service operation may mean modifying accustomed practices. Restoring community groundwater in Nashik, India, for example, required farmers' consensual commitment to using drip irrigation and new, low-energy water-pumping procedures.
• **Facilitate partnerships for coproduction.** Professionals, local authorities, and community members or civil society groups can, under shared governance, distribute

responsibility for planning and implementation. In Da Loc, Viet Nam, for example, CARE took responsibility for planning the restoration of mangrove coastal buffers, but the community led the establishment of a program that ensured access to shared resources.

Supporting Entrepreneurship

Without abdicating their responsibility for regulation or basic governance, local authorities that are unable to provide their populations with critical services will sometimes need to step aside, allowing motivated civic- or NGO-led groups to exercise initiative and creativity in implementing infrastructure services. For example, in the face of the government's failure to provide basic water and sanitation, Lufumbu, Tanzania, used locally appropriate technology to create its own water-collection and distribution systems. In addition to ensuring regular access to clean water, the effort improved public health, added a new cash crop, strengthened the status of women (by eliminating the burden of water collection), and ultimately gave rise to a transparent and democratic local governance system. Similarly, in Ixtlán, Mexico, indigenous residents financed their own infrastructure and created exemplary, cooperatively owned commercial enterprises. Such grassroots, culturally appropriate approaches offer valuable models for self-determination that should be encouraged.

To facilitate locally led infrastructural entrepreneurship, governments can provide authorization and timely regulatory review. And to augment the reach of self-funded initiatives and encourage replication, governments can provide incentives through tax relief, subsidies, or grants or awards for best practices. For their part, NGOs can provide communities with technical assistance, which is often sorely lacking even when projects have received funding support.

Haiti Redux: A "Future-Proof" Vision?

Many emerging economies have the potential to avoid the inefficiency and carbon intensity of industrial-era infrastructure and to move toward postindustrial solutions modeled on the closed-loop exchanges of natural systems. Given the general lack of infrastructure in the island nation of Haiti—with its crowded urban areas and utility-free hinterland—there are ample areas in which a tabula rasa approach might be feasible.[27]

Haiti's Caracol Industrial Park, introduced at the beginning of this book, was just such an attempt, but it was poorly conceived. Ultimately, Caracol not only displaced agricultural livelihoods and jeopardized a valuable marine conservation area but also largely failed to provide the local community with basic services—roadways, public transport, power, potable water, sewerage, drainage, and waste management. Although the park continues to operate (and to generate grand promises), it has largely perpetuated the intractable poverty of the area it was intended to serve.

Lacking both the capacity and capital to deliver infrastructural services, Haiti has relied on the piecemeal solutions offered by other nations and (especially) by NGOs in order to make recent small-scale, single-sector advances in wastewater treatment, solid waste collection, reforestation, and distributed solar power. As an alternative, however, Haiti might attempt next-generation, integrated solutions that are responsive to local context, provide critical services, and offer broad economic, social, and ecological benefits.

As part of an effort to stimulate the economy near the seaside village of Anse-Rouge (population 4,000)—which is located in the Artibonite Department, in the country's arid northwest region—the Haitian Ministry of Industry and Commerce put out a call inviting prospective developers to submit proposals for a new microindustrial park (such parks are characterized by smaller, more modest industrial clusters). A study developed in response proposes an ecoindustrial park (EIP) that would also provide critical services.[28] In addition to serving the EIP, the critical infrastructure included in the proposal would extend to residents of the village and beyond.

Anse-Rouge has virtually no infrastructure: sanitation is nonexistent, and water is scarce. Unemployment is high, and education levels are extremely low. The area has undergone significant environmental degradation, including major deforestation and consequently the loss of topsoil. Local food production is insufficient. The primary livelihood is salt farming. Laborers (mostly women) walk miles to salt-evaporation basins that are held by large landowners. There, the workers toil under a hot sun, harvesting sea salt. The salt is of poor quality and lacks sufficient natural iodine; however, because it is inexpensive, it is used widely throughout the country, contributing to widespread public health problems—including goiter and mental retardation—that are associated with iodine deficiency.

Because one of this proposal's key objectives is to foster the development of a locally restorative and regenerative circular economy—one that also fosters economic growth, social equity, and environmental responsibility—both the park and the services would be based on the principles of infrastructural ecology. The park's closed-loop system would eliminate waste, avert local pollution, and curb the drawdown of nonrenewable resources. The project would also incorporate strategies to support the environmental regeneration of the village's immediate surroundings. Finally, the proposal calls for creation of a public plaza, a small community center, and restrooms for the salt workers, all on the perimeter of the EIP.

The proposed anchor tenant of the EIP is a salt-refining factory, which would clean the local sea salt and fortify it with iodine. The design proposal also includes a dozen seawater greenhouses, which would produce food that could be processed in another of the EIP's enterprises, a commercial kitchen, which could be used to can or dry food that would then be sold to supplement local diets. The study estimates that excess water from the greenhouses might provide enough drinking water for the entire village, eliminating the need to truck in water for purchase. Potable domestic water, produced by the seawater greenhouses and adjacent solar stills (which use solar energy to evaporate, condense, and collect seawater), could be heated by solar thermal collectors and distributed to the EIP for food processing and other commercial uses. The salty brine rejected by the greenhouses would revert to the salt factory as input to the proposed cosmetic products, and the brine shrimp harvested from the salt basins could feed the fish farmed within the greenhouses' aquaponic systems, which combine conventional aquaculture with plant cultivation in a symbiotic, no-waste environment. The proposal includes more than a dozen such synergies (see figure 8.1).

The low hills adjacent to the site are separated by ravines, providing an opportunity to introduce water catchment for the collection of seasonal rains. The integrated water strategy proposed for the area immediately surrounding the EIP includes the construction of earthen surface impoundments (similar to the tank systems historically used in parts of India) to retain seasonal rains. To address both hydrologic and agricultural needs, the interconnected system of tanks would be integrated into newly constructed agricultural terraces—landscaped tiers that would rely on the principles of agroforestry to rebuild eroded soils, enabling them to retain moisture (see figure 8.2).

To meet the energy demands of most of the facilities (except the salt factory, which has been provisionally designed to be diesel powered), floating photovoltaic arrays would be placed above the water-catchment tanks, where they would reduce evaporative loss and inhibit algal growth—an arrangement that would carry the added advantage of keeping the solar panels cool, thereby increasing efficiency. Additional electricity, which would be distributed through a microgrid, would be generated by still more PV arrays, situated on building roofs and along the front face of the terraces' retaining walls.

For sanitation, both the EIP and the village would rely on a combination of centralized and distributed biodigesters, which would transform human, food, and agricultural waste from the immediate region into fertilizer and cooking gas (thereby reducing the use of charcoal). Because flood control is essential for this coastal area, the EIP would be carefully sited at an elevation well above storm-surge levels, in a location that would limit exposure to rainy-season mountain runoff (a risk that would, in any event, be minimized by the diversion of runoff to the agricultural terraces).

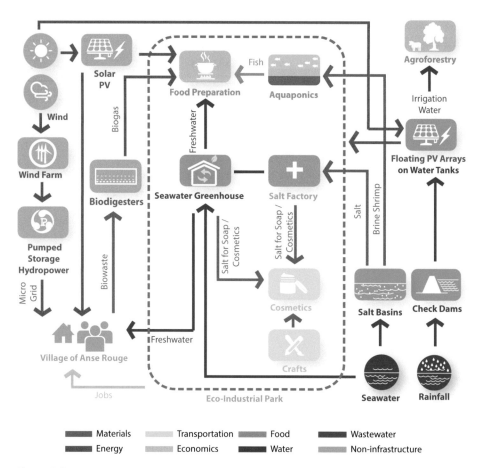

Figure 8.1
Energy, water, and resource flows, proposed ecoindustrial park, Anse-Rouge, Haiti. Credit: Hillary Brown and Logman Arja.

Beyond solar and biogas, the proposal includes a conceptual study of a more ambitious renewable-energy plan, a separate but related project that could potentially be undertaken, over the long term, by Electricité Haïti, the nation's electrical utility. If implemented, it could conceivably yield sufficient electrical power for a large portion of Haiti's northwest region. The plan calls for a wind-powered, pumped-storage hydroelectric system situated in the mountain slopes a few kilometers above Anse-Rouge, where reconditioned turbines (older-generation products cast off by the developed world) could be used to take advantage of the abundant trade winds (see figure 8.2). The accompanying reservoirs (one high, one low) would be carefully sited to benefit from

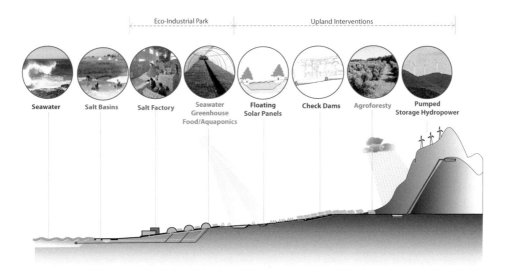

Figure 8.2
Section through proposed ecoindustrial park showing component relationships, Anse-Rouge, Haiti. Credit: Hillary Brown and Logman Arja.

existing topography, road access, and local springs, which could replenish evaporative loss from the reservoirs.

The EIP and its associated infrastructure—power, sanitation, waste management, and potable water—could significantly strengthen the area's economic viability: synergistic use of local resources (sun, sea, soil, and salt) could yield an estimated 200 to 300 new jobs, plus seasonal maintenance work.[29] The combination of employment and critical services should foster greater self-sufficiency, empower local residents, improve the lot of salt-basin workers, and increase local incomes. In other words, the proposal envisions a robust exemplar of the wide-ranging social, economic, and ecological benefits of infrastructural ecology.

Conclusion

Public works that adhere to the principles of infrastructural ecology are modeled on—and designed to work with—natural ecosystems. What David Orr has referred to as "the careful meshing of human purposes with the larger patterns and flows of the natural world" yields valuable gains in efficiency while reducing environmental impacts.[30] Implemented through collaborative, interdisciplinary leadership and an integrated body of policies, infrastructural ecology is a systemic innovation that contributes to green growth by moving beyond sector-specific thinking. It succeeds through simultaneous, cross-sector analyses of energy and resource flows, colocation of infrastructure assets, and new connections between—or redistribution of—waste energy and flows.

The goal of this book is to encourage emerging economies—and the investors and NGOs that work with them—to move toward more efficient, lower-carbon, climate-resilient, and context-sensitive infrastructure that capitalizes on opportunities for synergy between built systems and between built and natural systems. The case studies describe innovative services that have succeeded in both rural and urban locales—including settlements experiencing rapid immigration on top of natural population increase. The examples not only illuminate the potential adaptation and transferability of individual solutions but also highlight the potential to overcome the challenges enumerated in the introduction to the book. The book celebrates the many innovations that have successfully brought critical services to fledgling economies.

Summary of Case Study Infrastructural Ecologies

Chapter 2 - Case Study Summary Matrix

Project — Involved Sectors	Transportation	Energy	Water	Health Care	Telecom	Relational Solutions	Ecological Alignments
Jamuna Multi-purpose Bridge, Bangladesh	✓ (Car, Rail)	✓			✓	Conveyance of Electrical, Gas, and Telecom conduit and cables	
Khadju Bridge, Isfahan, Iran	✓	✓				Upstream irrigation	
Chinampas, Mexico	✓		✓	Sanitation & agriculture		Intercropping, Self-Irrigating, Self-fertilizing, Aqua-culture, Canal waterways	Erosion prevention
Multi Utility Tunnel, New Delhi, India		Multiple utilities in single tunnel				Electric, Gas, Water, Potable, Non-potable, Fiber-optics	
SMART Tunnel, Kuala Lumpur, Malaysia	✓		✓			Urban vehicular traffic diversion, Combined with, or replaced by, Stormwater storage	
ColaLife's AidPod, Zambia	✓			✓		Soft drink supply chain delivers medicine	
Vodafone's M-Pesa, Kenya and Beyond				Financial	✓	Remote population connected to banking services via cellphone	
Playpump, Gaviotas, Colombia		✓	✓			Well-water-pumping and water storage energized by play on children's seesaw	
EarthSpark's Micro-grid, Les Anglais, Haiti		✓ Electricity			✓	Micro-grid powered by surplus Digicel cell tower electricity	

Low-Carbon Process	Resilient Construction	Co-Development	Co-Benefits
			Capital savings, Resource conservation, Economic development
	Seasonal water management		Ceremonial, Social, and Commercial space
	Manages climate variability	Local labor	Micro-climate and intercropping improve crop yields
	Provision for future cooling water		Capital savings, Shared operating/maintenance expenses
	Response to increased monsoonal intensity		Capital savings ($1.58B), Productivity gains ($1.26B)
		Local labor	Reduced child mortality, Extended public health benefits
			Rural economic development
"Renewable" power resource		Local (child) labor through play	Avoided capital and operational electric costs for pumping
Capturing wasted electricity for local lighting and power		Local labor	Public safety

Chapter 2 - Case Study Summary Matrix

Project / Involved Sectors	Transportation	Energy	Water	Health Care	Telecom	Relational Solutions	Ecological Alignments
Solar-topped Narmada Branch Canal, Gujarat, India		✓	✓			Solar arrays reduce irrigation water loss, Water keeps PV panels cool, improving productivity/longevity	
		Electricty					
Trung Son Hydropower Multi-purpose Dam, Northwest Viet Nam		✓	✓			Flood control, Crop irrigation	
		Electricty					
Chinese Polyculture, (Dike-Pond System) Zhujiang Delta, China			✓			Flood control, aquaculture, Animal husbandry and silk production as part of a "circular economy"	
		Food production, Silk economy, with multiple reciprocities					
Tianjin Economic Technological Development Area, China		✓	✓			Water and energy cascading, Desalination, Wastewater treatment, Solid waste management	New soil from waste products, Pharmaceutical wastes as fertilizer
		Industrial park					

Low-Carbon Process	Resilient Construction	Co-Development	Co-Benefits

Low-Carbon Process	Resilient Construction	Co-Development	Co-Benefits
Renewable energy			Land conservation for agriculture, 15% lower capital and operation costs for energy production
Renewable energy			Avoided 1 million tons CO_2, New access roads
			Diverse diet for population
Recovered waste heat			Major CO_2 abatement, $17.6M revenue generation, Bottom ash as feedstock to new products

Chapter 3 - Case Study Summary Matrix

Project — Involved Sectors	Potable Water	Stormwater	Groundwater	Wastewater	Relational Solutions	Ecological Alignments
Sri Lanka Rainwater Harvesting and Storage	✓				Stormwater	Increased supply, Stormwater mitigation, Aquifer recharge
Restoration of India's "Tankas," Maner Basin		✓ Rainwater storage (groundwater) (irrigation)			Aquifer storage, Agriculture	Groundwater recharge, Flood control, Increased tank capacity
Curitiba, Brazil Flood Control		✓ Non-structural stormwater management			Urban parkland recreation, Urban amenity	Increased flood storage capacity, Increased biodiversity
Indonesia's "Biopori"		✓			Groundwater	Increased flood storage capacity
Humbo District, Ethiopia Reforestation	✓		✓		Replenishing aquifer	Erosion control, Improved micro-climate and biodiversity
East Kolkata Wetlands, India				✓	Aquaculture, Agriculture, MSW management	
Lesotho (Africa) Sewerage	✓			✓	Energy recovery of biogas, Sanitation	Extraction of biogas from wastewater, Improved sanitation
La Farfana WWTP, Santiago, Chile				✓	Energy recovery of biogas from wastewater	

Low-Carbon Process	Resilient Construction	Co-Development	Co-Benefits
30% reduction in water bills	Reduction in urban flooding	Community participation: local labor	Household time-savings, Improved public health
Pumping energy savings		Community participation: local labor	Local wages, Improved crop yields, Restored aquaculture, Nutrient cycles, Cost of fertilizer, Reduced out-migration
	Elimination of flood damages, 5x less cost than structural system		Factor 64 increase in per capita open space ratio, Improved local property values and public health
	Groundwater recharge, Flood control	Local labor	Beneficial fertilizer, Improved sanitation habits, Reduced land subsidence
		Local labor, Community cooperatives	Community revenue, Land titling, Revenue from timber sales, Carbon sequestration
Low-cost, Low-carbon wastewater treatment		Trad. knowledge preservation, Local labor	Ecosystem services valued at $38M, Rice, fish, and vegetable production
		Trad. construction knowledge, Local labor	Low-cost fertilizer, Reduced water pollution,1,500L/yr organic fertilizer
Elimination of abattoir, Biogas for 35,000 homes			CO_2 reduction of 2,634 tons/yr, CDM carbon credits

Chapter 4 - Case Study Summary Matrix

Project / Involved Sectors	Renewables	Comm. Heating	Transport	Wastewater	Telecom	Relational Solutions	Ecological Alignments
"iShack" Africa, Stellenbosch, South Africa	✓ Solar					Housing	
Micro-hydropower Nepal	✓ Hydro					Water	Energy from stream flow
Omnigrid Micropower Uttar Pradesh, India	✓				✓	Telecommuni-cations	
El Hierro Pumped Storage	✓ Wind					Energy storage, Desalination, Electric vehicles	Use of natural reservoir
Solar Hot Water Betim, Brazil		✓ Solar-heated domestic hot water					
Ouje-Bougoumou District Heating		✓ Community heating				Forestry (logging industry waste)	
Kigutu, Burundi Hospital				✓		Agriculture, Sanitation, Biogas cooking	Agroforestry
Idaban, Nigeria Biodigestion	✓					Animal husbandry	Cooking gas produced from biowaste

Low-Carbon Process	Resilient Construction	Co-Development	Co-Benefits
Housing with household energy and security, Low-cost home electricity		Local labor	Local employment in cottage industry, Tenure assistance, Rainwater collection
Household electricity access at low cost		Community co-funding and co-management	Reduction of deforestation, Household income genera-tion, New cottage industry
Off-grid telecom power, Low-cost village power and light		Local labor	Short-payback (6 mos.) and lower capital and operating costs (e.g., eliminated wiring), Improved village productivity
Reduction of 20,613 tons CO_2/yr			Island energy autonomy, 100% renewable power, High-efficiency battery
25% savings in residential electricity use		Local labor	Local jobs, Full municipal rollout of program
Shared community-wide savings, 2,300 tons CO_2/yr reduction 3.8 tons SOx 1.0 ton NOx		Largely community financed	District space heating and hot water, Reduced cost of heating
	Improved ground-water quality	Management training	Improved air quality, Fertilizer for gardens, Local jobs
Renewable cooking fuel		Local labor	.22 tons CO_2/yr reduced, Yields 1,500 liters/day organic fertilzer, Reduced water pollution

Chapter 4 - Case Study Summary Matrix

Project / Involved Sectors	Renewables	Comm. Heating	Transport	Wastewater	Telecom	Relational Solutions	Ecological Alignments
Belo Horizonte Landfill Gas-to-Energy	✓					Sanitation (MSW), Municipal composting, Recycling center	Extraction of biogas from wastewater, Improved sanitation
Itaipu Binacional Initiative, Ajuricaba, Brazil	✓					Agriculture, Animal husbandry, Agroforestry, Itapu Tech Park Foundation	Cleaning water for hydropower, Agroforestry from biomass and biodigestion
Volta Redonda "Fat to Fuel"			✓			Food industry	Cooking oil converted to biodiesel
DaNang, Viet Nam Environmental City	✓ (Hydro)			✓		Transportation, Housing	Stormwater management
Aerial Cable Cars in Latin America			✓			Community health	Access to ecological parks outside city
Solecshaw Solar Rickshaw in India			✓			Renewables, Air quality	Renewable energy sources

Low-Carbon Process	Resilient Construction	Co-Development	Co-Benefits
Carbon credits (CDM), 6% of value of biogas generated electricity		Landfill gas to electricity, Environmental education center	Revenue from sale of CERs, Community education, Elimination of methane emission
Agricultural methane elimination, Electrifies 2,200 local residents		Local labor, Organic farming	Thermal energy used in crop drying, Biodigester and cookstove production enterprises, Community revenue from gas sales
Reduction in particulate emissions		Local schools as collection facilities	Revenue stream for schools, Used oil residuals as animal feed
Walkable city planning for no-carbon transport	Reduction of flooding and traffic congestion	Community-engaged planning process	Flood mitigation, Tourism development
	Community connectivity and integration	Schools, Libraries co-located with stations	Access to employment
Locally generated solar power		Product develop-ment with rickshaw drivers	Local air quality benefits, Lower operating cost

Chapter 5 - Case Study Summary Matrix

Project / Involved Sectors	Desalination	Supply Irrigation	Groundwater	Wastewater	Defense	Relational Solutions	Ecological Alignments
Jakarta, Indonesia Coastal Defense					✓ (Sea wall)	Water storage enlargement, Raised transportation corridor, Urban parkland creation	Increased transpiration, Increased water storage
Red River Delta, Vietnam Mangrove Restoration					✓ (Coastal, Non-structural)	Mari-culture, Forestry	Ecosystem services restoration
Southwestern, Bangladesh Floating Gardens					✓ (Riverine)	Local resiliency, Agriculture, Animal husbandry	Multi-functional baira (floating agriculture beds), Fertilization
Makoko "Floating City," Lagos, Nigeria					✓ (Coastal)	Building adaptation, Self-sufficient floating facility "hot spots"	
Kibera Public Space Project, Nairobi, Kenya					✓ (River)	Wastewater	Stormwater management and restoration
Delhi's Urban Heat Island Policies				✓ (Buildings, Urban heat)		Community health, Air quality	Green roof requirements, Restoring air temperatures
Cambodia's Resilient Road Design Program				✓ (Roads and transport)		Agriculture	Stormwater restoration

Low-Carbon Process	Resilient Construction	Co-Development	Co-Benefits
	Flooding mitigation, Water storage, Multi-purpose construction		Recreational facilities, Subsidized replacement housing
$218M carbon reduction		Local labor, Disaster preparedness training	$15M value land and property protection, Fostering land building
	Stability of food production, Disaster preparedness	Indigenous activity	Increased agricultural productivity, International replication
Solar electricity	Community facilities and infrastructure provision	Participatory design, Community mobilization, Local labor	Improved public health, Local jobs
	Flood protection along N'gong River	Community engaged planning and training	Flood mitigation, Reduced health impacts of flooding
Green building codes, Reduced A/C energy use	Reduced heat impacts to community	Training programs and outreach	Air quality, Economic development
Locally generated solar power	Flood prevention and adaptation	Community participation in road building	Access to markets for cash crops

Chapter 5 - Case Study Summary Matrix

Project / Involved Sectors	Desalination	Supply Irrigation	Groundwater	Wastewater	Defense	Relational Solutions	Ecological Alignments
Restoration of "Amunas" Lima, Peru		✓				Animal husbandry	Hydrologic restoration, Grassland regeneration
Fergana Valley, Kazakhstan Water Banking		✓				Water banking, Increased hydropower production	Water rebalancing, Restoration of hydro dams
Jordan Valley Agriculture				✓		Agriculture	Treatment of reclaimed wastewater for crop irrigation mixed w/ grey water recovery of wastewater nutrients
Zimbabwe			✓			Controlled animal husbandry	400% increase in grass forage, 31% reduction in bare ground,17% decrease in soil loss (all in 1 year)
Hybrid Power Plant, Al Khafji, Saudi Arabia	✓					Renewable energy production	
Seawater greenhouse Qatar pilot	✓					Greenhouse agriculture, Mariculture, Algae production	Cooled and moist GH exhaust air increases exterior crop production

Low-Carbon Process	Resilient Construction	Co-Development	Co-Benefits
	Reduces water stress	Indigenous technique revived, New local grazing practices	Additional 40M m3/yr water availability, 31% cost savings over desalination
	Increased energy production does not limit water sufficiency		Irrigation water flows through local seasonal groundwater storage
Eliminated petroleum fertilizer	Reduces water stress	Farmer training in "fertigation"	$5M savings in fertilizer costs, Reduced soil salinization
	Pasture restored surface waters and pasture hydrologic restoration	Tribal leaders and local herders	International replication of process, Local land management training
Low carbon desalination			60,000 m³/yr of potable water
4/5 of distilled water production available for other uses			Interior cooling by 15°C algae, Bioenergy crops fish production

Chapter 6 - Case Study Summary Matrix

Project / Involved Sectors	Renewables	Grid Electricity	Water	Wastewater	Defense	Waste Mgmt	Relational Solutions	Ecological Alignments
Decentralizing Rural Viet Nam Infrastructure Development	Roads, Water systems, Public facilities						Local employment, Training and capacity building	
Jepirachi Wind Farm, Guarjira Peninsula, Colombia	✓						Protection/conservation of indigenous settlement	Community-identified environmental impacts mitigated
Mini-Grid, Casablanca, Morocco		✓					Slum electrification	
Urban NEXUS Program, Nashik, India			✓				Electricity, Peri-urban agriculture, Cooking gas (biogas) production	Drip-irrigation (conservation), Groundwater recharge
Mangrove Restoration, Da Loc, Viet Nam				✓			Forestry, Agriculture, Mariculture	Dyke maintenance, Mangrove planting and cleaning, Salination abatement
Dike and Sanitation & Dike Entrepreneurial Construction, Rufisque Senegal				✓	✓		Wetland and wastewater integration, Sanitation	Fertilizer production

Low-Carbon Process	Resilient Construction	Co-Development	Co-Benefits
		Consultation and coproduction, Local delegation of decision-making, Project selection, Local (in-kind) labor	Improvement of ties between rural communities and local authorities, Project selection, Local funding
Community benefit funds ($0.50/ton) CO_2 emission reduction credits		Colocation, Consultation and some decision-making	Healthcare upgrade, Sanitation, Renewably powered desalination plan, Local electrification, Job creation
€1.4M in reduced annual theft; 3-year payback to LYDEC		Community-representative operated grid	3,500 new jobs, Water and sewage connections, Program replication in other cities
20% electricity savings, Groundwater replenishment, Biodigestion reactivation		Labor for pumping and irrigation upgrades, Biodigester re-utilization	Organic farming training, Children's environmental education
	"Green team" waste management	Rebuilding coastal buffers, Mariculture stabilization, Commons mgmt.	Shared forestry yields, Program replication along coastline, Property protection, Improved rice production
Low-cost sanitation and wastewater treatment: 70% cost reduction, 56,000 tons of compost for fertilizer	Dike construction, Beach nourishment	Self-funded 100% construction and 70% maintenance labor	Improved public health job creation program replication: national and international

Chapter 6 - Case Study Summary Matrix

Project / Involved Sectors	Renewables	Grid Electricity	Water	Wastewater	Defense	Waste Mgmt	Relational Solutions	Ecological Alignments
Waste Management Implementation, Belo Horizonte, Brazil						✓	Construction waste handling, Recycling industry	Reduced pollution
Ixtlan de Juarez, Mexico Forestry Enterprise	✓						Forestry, Industry,	Ecosystem management, Biodiversity protection
Potable Water Provision, Lufumbu, Tanzania		✓ (Water potable)					Irrigation, Sanitation	Reforestation

Low-Carbon Process	Resilient Construction	Co-Development	Co-Benefits
		Entrepreneurship of local waste scavengers, Catholic Church	Worker "professionalization" and training, Worker health, Education, Wide program replication
	Production of (future) climate-adapted tree species	Local entrepreneurship and control of forest rights	Forest rights, SMART-WOOD certification, Critical infrastructure provision, Social and health benefits
		Local design and labor with techn. assist.	Elimination of water hauling, Improved health, Improved governance

Chapter 8 - Case Study Summary Matrix

Project Involved Sectors	Transportation · Energy · Water · Health Care · Telecom 	Relational Solutions	Ecological Alignments
Proposed Eco-Industrial Park, Anse-Rouge, Haiti	✓ ✓ Sanitation, Food production, Industrial activity	Seawater greenhouses yield food, fish, salt, PV-topped water bodies, Intercropping and agroforestry	Water catchment and infiltration, Erosion reduction, Gradual rebuilding of soil

Low-Carbon Process	*Resilient Construction*	*Co-Development*	*Co-Benefits*
Renewable energy, Energy storage	Responses to sea level rise, Increased aridity	Local labor intended	Potable water to community, Economic development, Local jobs

Notes

Chapter 1

1. United Nations Development Policy Analysis Division, LDC criteria data, 2015, http://www.un.org/en/development/desa/policy/cdp/ldc/ldc_data.shtml.

2. Amar Bhattacharya, Mattia Romani, and Nicholas Stern, *Infrastructure for Development: Meeting the Challenge* (London: Centre for Climate Change Economics and Policy, Grantham Research Institute on Climate Change and the Environment, and the Intergovernmental Group of Twenty Four, June 2012), 5.

3. Jacqueline Charles, "From Miami CAP Executive to Haiti Industrial Park Promoter," *Miami Herald*, May 6, 2015, http://www.miamiherald.com/news/nation-world/world/americas/haiti/article20380224.html.

4. Ayiti Kale Je Haiti Grassroots Watch, "Ayiti Veedor: Caracol Industrial Park (CIP): By the Numbers," http://haitigrassrootswatch.squarespace.com/28_eng.

5. Speech delivered by President Michel Martelly at the Caracol Industrial Park inauguration, November 28, 2011.

6. The plant was later expanded to 25 MW.

7. Haiti Grassroots Watch, "The Case of Caracol," January 2014, http://haitigrassrootswatch.squarespace.com/11_6_eng.

8. Ernesto M. Flores, Stefanie Brackmann, Jonathan Renshaw, Graham Watkins, Elizabeth Brito, Melissa Barandiaran, Crystal Fenwick, Maria Da Cunha, Tracey Friedlander, Suzanne Casolaro, and France François, "Environmental and Social Management Report (SMR), Haiti Productive Infrastructure Program II. Environmental Category: A, HA-L1081," Inter-American Development Bank, November 18, 2013, 1–69 at 5.

9. Ibid., 5.

10. Ibid., 7.

11. Jonathan M. Katz, "A Glittering Industrial Park in Haiti Falls Short," *Al Jazeera*, September 10, 2013, http://america.aljazeera.com/articles/2013/9/10/a-glittering-industrialparkfallsshortinhaiti.html.

12. Ironically, in December 2013, subsequent to the release of the IDB report, the government of Haiti announced that the bay would be one of three critical protected marine areas.

13. David Gootnick, director, international affairs and trade, United States Government Accountability Office, "Haiti Reconstruction: USAID Infrastructure Projects Have Had Mixed Results and Face Sustainability Challenges," October 9, 2013, GAO-13–558, June 2013, 50, http://www.gao.gov/assets/660/655278.pdf.

14. U.S. Agency for International Development (USAID), "Caracol Industrial Park," https://www.usaid.gov/haiti/caracol-industrial-park.

15. Kysseline Jean-Mary Chérestalm, "Building Back Better?: The Caracol Industrial Park and Post-earthquake Aid to Haiti," ActionAid USA, January 2015, 13, http://www.actionaidusa.org/sites/files/actionaid/building_back_better_the_caracol_industrial_park_and_post-earthquake_aid_to_haiti.pdf.

16. Meg Findley and Martin Côté, AECOM, "Environmental Assessment of the USAID/Haiti North Park Power Project," *USAID* (June 2011), 1–126 at 8, http://www.ute.gouv.ht/caracol/images/stories/docs/environmental%20assessment%20of%20the%20usaidhaiti%20north%20park%20power%20project.pdf.

17. Ibid.

18. Center for Economic and Policy Research, "Second USAID Contractor Suspended Following Caracol Housing Debacle," March 30, 2015, http://cepr.net/blogs/haiti-relief-and-reconstruction-watch/second-usaid-contractor-suspended-following-caracol-housing-debacle.

19. Hillary Brown, *Next Generation Infrastructure*: *Principles for Post-Industrial Public Works* (Washington, DC: Island Press, 2014), focuses predominantly on North America, Northern Europe, and Japan. As is the case in this volume, the guiding principles for the development of integrated, holistic projects were drawn largely from the fields of industrial symbiosis or industrial ecology.

20. United Nations, Department of Economic and Social Affairs, *World Economic and Social Survey* (New York: United Nations, 2013).

21. United Nations, Department of Economic and Social Affairs, Population Division, *World Urbanization Prospects: The 2014 Revision, Highlights*, ST/ESA/SER.A/352 (New York: United Nations, 2014).

22. V. Henderson, "Urbanization in Developing Countries," *The World Bank Research Observer* 17, no. 1 (2002): 89–112. doi:10.1093/wbro/17.1.89.

23. United Nations, Department of Economic and Social Affairs, Population Division, *World Urbanization Prospects*. During those same years, Canada's urbanization rate was 0.2 percent.

24. United Nations Human Settlements Program (UN Habitat), *State of the World's Cities 2010/2011: Bridging the Urban Divide* (London: Earthscan, 2008).

25. The World Bank: The International Bank for Reconstruction and Development, *Global Monitoring Report 2013: Rural-Urban Dynamics and the Millennium Development Goals* (Washington, DC: World Bank, 2013).

26. The World Bank: The International Bank for Reconstruction and Development, *Infrastructure at the Crossroads: Lessons from 20 Years of World Bank Experience* (Washington, DC: World Bank, 2006), 48.

27. Ibid., 49.

28. Ibid., 81.

29. Lorna Dwyer, "Biopiracy, Trade and Sustainable Development," *Colorado Journal of Environmental Law and Policy* 19.2 (2008): 219–258.

30. M. L. Ross, "The Political Economy of the Resource Curse," *World Politics* 51, no. 2 (1999): 297–322. doi:10.1017/S0043887100008200.

31. Marian Miller, *The Third World in Global Environmental Politics* (London: Rienner, 1995), 31.

32. Ross, "Resource Curse."

33. Per Bolund and Sven Hunhammar, "Ecosystem Services in Urban Areas," *Ecological Economics* 29, no. 2 (1999): 293–301. doi:10.1016/S0921-8009(99)00013-0.

34. Ian Bannon and Paul Collier, eds., *Natural Resources and Violent Conflict: Options and Actions* (Washington, DC: World Bank Publications, 2003).

35. Michael Gallis, Gary Moll, and Heather Millar, "People-Nature: The Human Network," ESRI, 2007, http://www.esri.com/news/arcnews/fall07articles/people-nature.html.

36. Among the many publications on this topic, see Charlene Spretnak, *Relational Reality: New Discoveries of Interrelatedness That Are Transforming the Modern World* (Topsham, ME: Green Horizon Books/Green Horizon Foundation, 2011); Charlene Spretnak, *Resurgence of the Real: Body, Nature and Place in a Hypermodern World* (New York: Routledge, 1999); Chet Bowers, *The Culture of Denial: Why the Environmental Movement Needs a Strategy for Reforming Universities and Public Schools* (Albany: State University of New York Press, 1997).

37. Miller, *The Third World*, 27.

38. Brown, *Next Generation Infrastructure*.

39. Jouni Korhonen, "Industrial Ecology for Sustainable Development: Six Controversies in Theory Building," *Environmental Values* 14 (2005): 85. doi:10.3197/0963271053306096.

40. Janet A. M. Eilering and Walter J. V. Vermeulen, "Eco-Industrial Parks: Toward Industrial Symbiosis and Utility Sharing in Practice," *Progress in Industrial Ecology* 1, nos. 1–3 (2004): 245–270. doi:10.1504/PIE.2004.004681.

41. As applied here and in the chapters that follow, the term "colocation"—the placement of several entities in a single location—describes the physical arrangements of diverse infrastructural assets in shared locations or otherwise situated in close proximity.

42. According to experts in industrial ecology, in order to qualify as industrial symbiosis, at least three different entities must share at least two different resources.

43. Marian R. Chertow, "Uncovering Industrial Symbiosis," *Journal of Industrial Ecology* 11, no. 1 (2007): 12. doi:10.1162/jiec.2007.1110.

44. David Batten, "On the Dynamics of Industrial Evolution," *Regional Science and Urban Economics* 12, no. 3 (1982): 449–462. doi:10.1016/0166-0462(82)90029-1, 14.

45. Bryna Cosgriff Dunn and Anne Steinemann, "Industrial Ecology for Sustainable Communities," *Journal of Environmental Planning and Management* 41, no. 6 (1998): 661–672. doi:10.1080/09640569811353, 14.

46. Kalundborg Symbiosis, http://www.symbiosis.dk/en.

47. Teresa Domenech and Michael Davies, "Structure and Morphology of Industrial Symbiosis Networks: The Case of Kalundborg," *Procedia Social and Behavioral Sciences* 10 (2011): 79–89. doi:10.1016/j.sbspro.2011.01.011.

48. City of Stockholm, "Hammarby Sjostad Environmental Program" (Stockholm: City Planning Administration, City of Stockholm, 1996a), 1–10 at 4.

49. K. E. Eriksson and K. H. Robèrt, "From the Big Bang to Sustainable Societies," *Reviews in Oncology* 4, no. 2 (1991): 5–14.

50. Sofie Pandis Iveroth and Nils Brandt, "The Development of a Sustainable Urban District in Hammarby Sjöstad, Stockholm, Sweden?" *Environment, Development and Sustainability* 13 (2011): 1043. doi:10.1007/sl0668-011-9304-x.

51. Örjan Svane, Josefin Wangle, Lars A. Engberg, and Jenny Palm, "Compromise and Learning When Negotiating Sustainabilities: The Brownfield Development of Hammarby Sjöstad, Stockholm," *International Journal of Urban Sustainable Development* 3, no. 2 (November 2011): 141–155.

52. Sofie Pandis Iveroth, Anne-Lorène Vernay, Karel F. Mulder, and Nils Brandt, "Implications of Systems Integration at the Urban Level: The Case of Hammarby Sjöstad, Stockholm," *Journal of Cleaner Production* 48 (2013): 220–231.

53. Ibid., 224.

54. Ibid.

55. Sofie Pandis Iveroth, Stephan Johannson, and Nils Brandt, "The Potential of the Infrastructural Systems of Hammarby Sjöstad in Stockholm, Sweden," *Energy Policy* 59 (2013): 716–726.

56. Ibid., 724.

57. Sofie Pandis Iveroth and Nils Brandt, "Evaluation of Hammarby Sjöstad's Environmental Profiling," TRITA IM 2009: 03, Stockholm, Division of Industrial Ecology.

58. United Nations, Department of Economic and Social Affairs, Population Division, *World Urbanization Prospects: The 2014 Revision.*

Chapter 2

1. The World Bank, "Jamuna Bridge—A Boost for Bangladesh's Economy," *World Bank—Transport in South Asia*, http://go.worldbank.org/I4JRJD65V0.

2. In Europe and Asia, railroads, electricity transmission, and broadband corridors are commonly accommodated in the same right-of-way, providing efficiencies and cross-subsidies between these uses.

3. Fabio Galli, "Implementation Completion Report (IDA-25690) for the People's Republic of Bangladesh for the Jamuna Bridge Project," *The World Bank*, June 19, 2000, 10.

4. G. R. Kuros and E. L. Koneiki, "Water and Irrigation Techniques in Ancient Iran," Iranian National Committee on Irrigation and Drainage, May 2007.

5. Ibid.

6. Hubert Chanson, "Historical Development of Stepped Cascades for the Dissipation of Hydraulic Energy," *Transactions of the Newcomen Society* 72 (2000): 295–318.

7. This case study by Hillary Brown was previously published in the article: "Compound Infrastructure: Lessons from Engineering Precedents," *Journal of Bridge Engineering*, 19, no. 9 (September 2014): 02514001-1-6. doi:10.1061/(ASCE)BE.1943-5592.0000648. It is included here with permission from ASCE.

8. B. L. Turner II, "Prehistoric Intensive Agriculture in the Mayan Lowlands," *Science* 185 (July 12, 1974): 118–124 at 121.

9. Population estimated between 12,500 and 16,000 per square kilometer (50 and 60 per acre). See William Denevan, *The Native Population of the Americas in 1492*, second edition (Madison: University of Wisconsin Press, 1992), 148.

10. Pablo Torres-Lima, Beatrix Canabal-Cristiani, and Gilberto Burela-Rueda, "Urban Sustainable Agriculture: The Paradox of the Chinampa System in Mexico City," *Agriculture and Human Values* 11.1 (1994): 37–46 at 38. Recent studies are questioning the existence of these perimeter trees because there is insufficient archaeological evidence of their deployment.

11. V. Popper, "Investigating Chinampa Farming," *Back Dirt: Newsletter of the Institute of Archaeology, University of California, Los Angeles*, Fall/Winter 2000, 4–5.

12. Lima et al., "Urban Sustainable Agriculture," 39.

13. Ibid.

14. D. Renard, J. Iriarte, J. J. Birk, S. Rostain, B. Glaser, and D. McKey, "Ecological Engineers Ahead of Their Time: The Functioning of Pre-Columbian Raised-Field Agriculture and Its Potential Contributions to Sustainability Today," *Ecological Engineering* 45 (2012): 30–44 at 34.

15. Gregory T. Cushman, *Guano and the Opening of the Pacific World: A Global Ecological History* (Cambridge: Cambridge University Press, 2013), 283.

16. William M. Denevan, *Cultivated Landscapes of Native Amazonia and the Andes* (Oxford: Oxford University Press, 2001), 220.

17. Renard et al., "Ecological Engineers," 31.

18. Ibid., 34.

19. Charles H. Coe, "The Floating Gardens of Mexico (1895)," *Arnoldia* 60, no. 3 (2000): 44–46 at 44, http://arnoldia.arboretum.harvard.edu/pdf/articles/2000-60-3-the-floating-gardens-of-mexico -1895.pdf.

20. Ibid.

21. Denevan, *Cultivated Landscapes*, 236.

22. Axel Laistner and Hermann Laistner, "Utility Tunnels–Proven Sustainability Above and Below Ground," 2012, http://www.corp.at; Road Traffic-Technology, "SMART (Stormwater Management and Road Tunnel), Kuala Lumpur, Malaysia," 2013, http://www.roadtraffic-technology.com/projects /smart/.

23. "Marina Bay, Singapore Common Services Tunnel and District Cooling System," China Trend Building Press Limited, June 2006, http://www.building.hk/forum/2007_0309marinabay.pdf.

24. Gaurav Bhatnagar, "CP Tunnel: A Technological Feat That Went Unsung," *The Hindu* (New Delhi), July 23, 2013.

25. Rozana Sani, "Putrjaya—Ecofriendly Tech City," Vertilite Corporation, http://www.vertilite .asia/putrajaya-ecofriendly-tech-city.

26. Association for Progressive Communications News, "Developing Countries Helped to Achieve Universal Access through Use of Shared Infrastructure," April 15, 2015, https://www.apc.org/en /news/developing-countries-helped-achieve-universal-access.

27. D. V. L. Hunt, N. K. Drinkwater, and C. D. F. Rogers, "Sustainable Utility Placement for University Campuses," in *GeoCongress 2012: State of the Art and Practice in Geotechnical Engineering*, ed. Hryciw, Roman D., Adda Athanasopoulos-Zekkos, and Nazli Yesiller (Reston, VA: ASCE, 2012), 4309–4318.

28. Laistner and Laistner, "Utility Tunnels."

29. D. V. L. Hunt, D. Nash, and C. D. F. Rogers, "Sustainable Utility Placement via Multi-utility Tunnels," *Tunneling and Underground Space Technology* 39, no. 1 (2014): 15–26.

30. "Frequently Asked Questions," Laman Web Rasmi Pusat Kawalan SMART, http://state.water .gov.my/scc/index.php/en/faq.

31. Ibid.

32. "Success of Kuala Lumpur's Dual Purpose Tunnel," ITS International, July–August 2012, http://www.itsinternational.com/categories/detection-monitoring-machine-vision/features/success-of-kuala-lumpurs-dual-purpose-tunnel/.

33. G. Kados and Y. H. Kok, "Stormwater Management and Road Tunnel (SMART)," in *Underground Space—The 4th Dimension of Metropolises*, ed. Jirí Barták, Ivan Hrdina, Georgij Romancov, Jaromír Zlámal (London: Taylor and Francis, 2007), 1183.

34. Deti Mega Purnamasari, "Jokowi to Turn to Private Investors for Jakarta Tunnel," *Jakarta Globe*, May 8, 2013.

35. ColaLife, www.colalife.org.

36. As of December 2015, the kits are now available in 26 Shoprite stores nationwide. See ColaLife, http://www.colalife.org/2015/11/27/kit-yamoyo-launches-in-shoprite-stores-nationwide-in-zambia/.

37. World Bank, Financial and Private Development, http://web.worldbank.org/WBSITE/EXTERNAL/TOPICS/EXTFINANCIALSECTOR/0,,contentMDK:23174014~pagePK:210058~piPK:210062~theSitePK:282885,00.html.

38. Christine Lagarde, "Empowerment through Financial Inclusion," address to the International Forum for Financial Inclusion, Mexico, June 26, 2014, https://www.imf.org/external/np/speeches/2014/062614a.htm.

39. William Britt Gwinner, Michael J. Goldberg, Tova Maria Solo, and Alberto Didoni, "From Financial Exclusion to Inclusion: Increasing the Availability of Credit to the Urban Poor in Latin America," *En Breve* 77 (2005): 1–4.

40. Daniel Thomas and Katrina Mason, "Africa's Digital Money Heads to Europe," *Financial Times*, March 30, 2014; K. McKemey, N. Scott, D. Souter, T. Afullo, R. Kibombo, O. Sakyi-Dawson, "Innovative Demand Models for Telecommunications Services," Final Technical Report, Contract Number R8069, Department for International Development (DFID), 2003.

41. Jude Weber, "Developing World Blazes Trail for 2.5 bn with No Banking Access," *Financial Times*, http://www.ft.com/intl/cms/s/0/efffb19c-fdf2-11e3-bd0e-00144feab7de.html#axzz3geOLATts.

42. Thomas and Mason, "Africa's Digital Money."

43. Irene Mondago, "Out of Africa: Mobile Phone Banking Surges in EM," Beyondbrics blog, *Financial Times*, http://blogs.ft.com/beyond-brics/2014/07/07/out-of-africa-mobile-phone-banking-surges-in-em/.

44. Thomas and Mason, "Africa's Digital Money."

45. For further details about one of these technologies, see chapter 5.

46. A parafunctional object is one that is used in a way other than the most common use—in this instance, a seesaw is used as a pump.

47. This number was reported in 2003, so there are likely more in place. See Monica del Pilar Uribe Marin, "Times for Utopia," *New Internationalist*, no. 357 (June 2003).

48. Amy Costello, producer and reporter, "Southern Africa: Troubled Water," Public Broadcasting System, *Frontline/World*, July 29, 2010, http://www.pbs.org/frontlineworld/stories/southernafrica904 /video_index.html.

49. Morning Washburn, "Wanted: Clean, Smart, Affordable Energy for Haitian Community," *Frontlines Online Edition*, USAID, 2014, https://www.usaid.gov/news-information/frontlines/grand -challenges/wanted-clean-smart-affordable-energy-haitian-community.

50. Allison Archimbault, "EarthSpark International Community Engagement Process for the Development of Its Microgrid in Les Anglais, Haiti," White Paper, EarthSpark, September 11, 2012.

51. Microgrid designed and furnished with the help of Daniel C. Gregory, CEO of Positive Energies, LLC.

52. See chapter 5.

53. World Bank, "Access to Electricity (% of Population)," http://data.worldbank.org/indicator /EG.ELC.ACCS.ZS.

54. Anoop Singh, "A Policy for Improving Efficiency of Agriculture Pump Sets in India: Drivers, Barriers, and Indicators," *Climate Strategies*, UK, Working Paper, International Support for Domestic Action, 2009, http://climatestrategies.org/wp-content/uploads/2009/09/isda-indian-power-sector -21september-2009-report.pdf.

55. Government of India, Ministry of New and Renewable Energy, http://www.mnre.gov.in /schemes/grid-connected/solar/.

56. "Government Raises Solar Power Target to 100,000 MW by 2022," *Hindu Times*, http://www .thehindu.com/news/national/union-government-raises-solar-power-target-5times-to-100000 -mw-by-2022/article7325821.ece.

57. Pradeep S. Mehta and Smriti Bahety, "India's Solar Panel Dispute: A Need to Look Within," *Wire*, March 15, 2015, http://thewire.in/24787/indias-solar-panel-dispute-a-need-to-look-within/.

58. Government of India, Ministry of New and Renewable Energy, "Physical Progress (Achievements)," http://www.mnre.gov.in/mission-and-vision-2/achievements/.

59. Utpal Bhaskar, "Budget Stresses on Clean Energy Sources," *Hindustan Times*, July 10, 2014, http://www.livemint.com/Politics/fUzkHkQ5VKZ73v6KeoJpAM/Arun-Jaitley-allocates-Rs-1000 -crore-for-renewable-sector-s.html.

60. Beverly Mitchell, "India to Build World's Largest Onshore Floating Solar Power Project," Inhabitat, July 4, 2014, http://inhabitat.com/india-to-build-worlds-largest-onshore-floating-solar -power-project/.

61. Ibid.

62. Ucilia Wang, "A Solar Canal Rises in India," *Gigacom Research*, Tech News and Analysis, April 23, 2012, https://gigacom.com/2012/04/23/a-solar-canal-rises-in-india/.

63. Young-Kwon Choi, Nam-Hyung Lee, and Kern-Joong Kim, *Advances in Information Technology and Computer Science*, CES-CUBE 25 (2013): 284–289.

64. Sumit Khanna, "SSNNL to Set Up 10-MW Solar Power Project," August 1, 2013, http://www.dnaindia.com/ahmedaba/1868570/report-ssnnl-to-set-up-10mw-solar-power-projet.

65. Smiti Mittal, "Kyocera to Develop Japan's Largest Floating Solar Power Plant," CleanTechnica, September 22, 2014, http://cleantechnica.com/2014/09/22/kyocera-develop-japans-largest-floating-solar-power-plant/.

66. Richard Heasman, "India Plans to Build 'World's Largest' Floating Solar Power Plant," July 4, 2014, http://blueandgreentomorrow.com/2014/07/04/india-plans-to-build-worlds-largest-floating-solar-power-plant/.

67. Ibid.

68. UKAID, Department for International Development (DFID), "Water Storage and Hydropower: Supporting Growth, Resilience and Low Carbon Development," 2008, 14.

69. Trung Son Hydropower Company Limited, www.trungsonhp.vn/en/511-loi-ich-cua-du-an.html.

70. World Bank, "Trung Son Hydropower Project: Meeting Growing Energy Demands," *World Bank News*, November 20, 2012, http://www.worldbank.org/en/news/feature/2012/11/20/trung-son-hydropower-project-meeting-growing-energy-demands.

71. Hydro4Africa, "GEF Approves Mini-Hydro Infrastructure Project in Liberia," May 1, 2012, http://hydro4africa.net/news/viewnews.php?ID=676.

72. Man-ki Kim, "Korea Building World's Largest Tidal Power Plant," *Korean Herald*, March 30, 2010.

73. Vincent Denis, "Wastewater Turbining before and after Treatment: The Example of Amman City, Hashimite Kingdom of Jordan," Proceedings of the International Conference "Hydro 2007," Granada, Spain, October 15–17, 2007, 9.

74. ESHA, Mhylab, Acconia Energy, Dolas, IMP PAN, IWHW, Studio Frozio, WMF, Wasser Tirol Small Hydro Action for the Promotion of Efficient Solutions, "Infrastructures with Small Hydropower Plants: Multipurpose Schemes," The European Small Hydropower Association, August 29, 2011, http://www.esha.be/fileadmin/esha_files/documents/SHAPES/Multipurpose%20schemes%20brochure%20SHAPES.pdf.

75. Aline Choulot, Vincent Denis, and Petras Punys, "Integration of Small Hydro Turbines into Existing Water Infrastructure," in *Hydropower—Practice and Application*, ed. Hossein Samadi-Boroujeni (Rijeka, Croatia: INTECH, 2014).

76. Herbert Girardet, *The Gaia Atlas of Cities: New Directions for Sustainable Urban Living* (London: Gaia Books, 1992).

77. George L. Chan, "Aquaculture, Ecological Engineering: Lessons from China," *Ambio* 22, no. 7 (November 1993): 491–494.

78. K. Ruddle, J. I. Furtado, G. F Zhong, and H. Z. Deng, "The Mulberry Dike-carp Pond Resource System of the Zhujiang (Pearl River) Delta, People's Republic of China," *Applied Geography* (1983): 45–62 at 50.

79. Rick Parker, *Aquaculture Science*, second edition (Albany, NY: Delmar Press, 2002), 6.

80. Ibid.

81. Mae-Wan Ho, "Circular Economy of the Dyke-Pond System," Institute of Science in Society Report, September 10, 2006, http://www.i-sis.org.uk/DykePondSystem.php.

82. Han Shi, Marian Chertow, and Yuyan Song, "Developing Country Experience with Eco-industrial Parks: A Case Study of the Tianjin Economic-Technological Development Area," *Journal of Cleaner Production* 18 (2010): 191–199. The pilot project was not the first indication of TEDA's economic and environmental proclivities. In 2000, TEDA's administration had obtained ISO 14001 certification in order to improve its public image and gain a competitive advantage in the international market. See H. M. Wei, "Development of a Regional ISO 14001 Environmental Management System in TEDA," in *Proceedings of Sino-Japanese Workshop on Circular Economy and Zero Emission, Tianjin*, 2001.

83. Shi et al., "Developing Country Experience with Eco-industrial Parks," 195.

84. Yong Geng and Brent Doberstein, "Developing the Circular Economy in China: Challenges and Opportunities for Achieving 'Leapfrog' Development," *International Journal of Sustainable Development and World Ecology* 15 (2008): 231–239.

85. Shi et al., "Developing Country Experience with Eco-industrial Parks."

86. Ibid.

87. Ibid., 194.

88. Ibid., 196.

89. Lei Shi and Bing Yu, "Eco-industrial Parks from Strategic Niches to Development Mainstream: The Cases of China," *Sustainability* 2 (2014): 6325–6331. doi:10.3390/su6096325.

90. Shi et al., "Developing Country Experience with Eco-industrial Parks," 196.

91. Shi and Yu, "Eco-industrial Parks from Strategic Niches to Development Mainstream," 6331.

92. Zengwei Yuan, Jun Bi, and Yuichi Moriguichi, "The Circular Economy: A New Development Strategy in China," *Journal of Industrial Ecology* 10, nos. 1–2 (2006): 4–8.

93. Shi and Yu, "Eco-industrial Parks from Strategic Niches to Development Mainstream," 6326.

94. Yuan et al., "Circular Economy," 6.

95. May Hald, "Sustainable Urban Development and Chinese Eco-city: Concepts, Strategies, Policies and Assessments," *Fridtjof Nansen Institute Report* 5 (2009): 1–84.

Chapter 3

1. Cūlavaṃsa, LXVIII, 8 (Sri Lankan Historical chronicle). Translation by William Geiger.

2. The nexus of water, energy, and food production will be discussed in more detail in chapter 5.

3. Peter H. Gleick, "Global Freshwater Resources: Soft-Path Solutions for the 21st Century," *Science* 302, no. 5650 (November 14–28, 2003): 15324–15328. doi:10.1126/science.1089967.

4. A rain shadow is a dry area on the downwind side of a mountain.

5. Frances Murray and David Little, "The Nature of Small-Scale Farmer Managed Irrigation Systems in North West Province, Sri Lanka and Potential for Aquaculture," Working Paper SL 1.3., UK Department for International Development (DFID) funded natural resources research program R7064, 2000.

6. Madduma Bandara, "Catchment Ecosystems and Village Tank Cascades in the Dry Zone of Sri Lanka: A Time-Tested System of Land and Water Resource Management," in *Strategies for River Basin Management*, ed. Jan Lundqvist, Ulrik Lohm, and Malin Falkenmark (Netherlands: Springer, 1985), 99–113.

7. Ibid.

8. Murray and Little, "Small-Scale Farmer Managed Irrigation Systems," 8.

9. Ibid., 3.

10. Arnold Pacey, *The Maze of Ingenuity: Ideas and Idealism in the Development of Technology*, second edition (Cambridge, MA: MIT Press, 1992), 12–13.

11. Udo Wiesmann, In Su Choi, and Eva-Maria Dombrowski, *Fundamentals of Biological Wastewater Treatment* (Weinheim: Wiley-VCH Verlag, 2007), 4.

12. The revived interest in rainwater harvesting may result in part from the fact that rainwater can be of higher quality than surface or subgrade water. See Raquel Pinderhughes, *Alternative Urban Futures: Planning for Sustainable Development in Cities throughout the World* (Lanham, MD: Rowman and Littlefield, 2006), 37.

13. Stockholm Environment Institute, "Rainwater Harvesting: A Lifeline for Human Well-Being," United Nations Environment Programme and Stockholm Environment Institute, 2009, http://www.unwater.org/downloads/Rainwater_Harvesting_090310b.pdf.

14. Ibid., 47.

15. Government of India, Press Information Bureau, "India Cities Focus on Rainwater Harvesting to Provide Clean Drinking Water," January 4, 2010, http://www.circleofblue.org/waternews/2010/world/india-cities-focus-on-rainwater-harvesting-to-provide-clean-drinking-water/.

16. Ibid.

17. Jutta Schwengsbier, "Islamabad to Harvest Rain Water," *Deutsche Welle* (online), December 7, 2010, www.dw.com/en/Islamabad-to-harvest-rain-water/a-5787215.

18. Murray and Little, "Small-Scale Farmer Managed Irrigation Systems," 4.

19. Tanuja Ariyanada, "Domestic Rainwater Harvesting as a Water Supply Option in Sri Lanka," *Hydro Nepal* 6 (2010): 28.

20. Passanna Gunasekera, "For Rural Sri Lanka, Ancient Technology Eases Water Woes," USAID, *Frontlines Energy/Infrastructure* (January–February 2014): 1–5, https://www.usaid.gov/news-information /frontlines/energy-infrastructure/rural-sri-lanka-ancient-technology-eases-water-woes.

21. Ariyanada, "Domestic Rainwater," 29.

22. T. Ariyananda, S. S. Wickramasuiya, and D. S. Wijeyesekerari, "Rain Water Harvesting for Water Efficiency and Management," in *Proceedings: International Conference on Sustainable Built Environment*, Kandy, December 2010, 156–159.

23. Ibid.

24. Gunasekera, "Ancient Technology Eases Water Woes," 2.

25. Rakesh Kumar, R. D. Singh, and K. D. Sharma, "Water Resources of India," *Current Science* 89 (2005): 794.

26. Jamie Pittock, ed., "Water for Life: Lessons for Climate Change Adaptation from Better Management of Rivers for People and Nature," World Wildlife Fund, 2008, 18, http://awsassets.panda.org /downloads/50_12_wwf_climate_change_v2_full_report.pdf.

27. Ibid.

28. Biksham Gujja, Sraban Dalai, Hajara Shaik, and Vinod Goud, "Adapting to Climate Change in the Godavari River Basin of India by Restoring Traditional Water Storage Systems," *Climate and Development* 1 (2009): 229–240.

29. O. Aubriot, and P. I. Prabhakar, "Water Institutions and the 'Revival' of Tanks in South India: What Is at Stake Locally?" *Water Alternatives* 4, no. 3 (2011): 325–346 at 331.

30. Gujja et al., "Adapting to Climate Change," 232.

31. J. B. S. Umanadh, "On a Mission Mode to Save Tanks in Telangana," *Deccan Herald*, September 24, 2015.

32. K. Lenin Babu and S. Manasi, "Estimation of Ecosystem Services of Rejuvenated Irrigation Tanks: A Case Study in Mid Godavari Basin," Proceedings of the IWMI-TATA Water Policy Research Program "Managing Water in the Face of Growing Scarcity, Inequity and Declining Returns: Exploring Fresh Approaches" (Patancheru, Hyderabad, India: ICRISAT, 2008), 282–297 at 283.

33. Umanadh, "Save Tanks in Telangana."

34. Gujja et al., "Adapting to Climate Change," 232.

35. Babu and Manasi, "Estimation of Ecosystem Services," 284.

36. ManaPSC, "All You Need to Know about 'Mission Kakatiya,'" September 15, 2015, http://www
.manapsc.com/all-you-need-to-know-about-mission-kakatiya/.

37. S. Gadgil, "Climate Change and Agriculture—An Indian Perspective," *Current Science* 69, no. 8
(1998): 649–659.

38. Gujja et al., "Adapting to Climate Change," 237.

39. Ibid., 230.

40. Ibid., 233.

41. Pittock, "Water for Life," 13.

42. Gujja et al., "Adapting to Climate Change," 233–235.

43. Babu and Manasi, "Estimation of Ecosystem Services," 290.

44. Gujja et al., "Adapting to Climate Change," 235.

45. Babu and Manasi, "Estimation of Ecosystem Services," 286.

46. Ibid.

47. Ibid., 287.

48. Gujja et al., "Adapting to Climate Change," 235.

49. Babu and Manasi, "Estimation of Ecosystem Services," 292.

50. Nilesh Heda, "Conservation of Riverine Resources through People's Participation: North-
Eastern Godavari Basin Maharashtra, India," final report, Rufford Small Grant for Nature Conservation (Maharashtra: Samvardhan, 2011), 14.

51. Gujja et al., "Adapting to Climate Change," 233.

52. Pittock, "Water for Life," 18.

53. Umanadh, "Save Tanks in Telangana."

54. Ibid.

55. Aubriot and Prabhakar, "Water Institutions and the 'Revival' of Tanks in South India," 325.

56. S. Cairncross and E. A. R. Ouano, *Surface Water Drainage for Low-Income Communities* (Geneva:
World Health Organization, 1991).

57. Many recently adopted urban strategies, such as green roofs and permeable pavement, are also
classified as green infrastructure.

58. Carlos E. M. Tucci, "Brazil: Flood Management in Curitiba Metropolitan Area," Meteorological Organization/Global Water Partnership Associated Programme on Flood Management, http://
www.apfm.info/publications/casestudies/cs_brazil_full.pdf.

59. Statistics meaning that in any given year there is a 0.5 percent or 1 percent chance that a flood-prone area will flood.

60. Hanna-Ruth Gustafsson and Elizabeth Kelly, "Urban Innovations in Curitiba: A Case Study," Eugene and Carol Ludwig Center for Community and Economic Development, *Yale Law School*, 2012, https://www.law.yale.edu/documents/pdf/News_&_Events/LudwigGustafssonKellyCuritiba Report.pdf.

61. Tucci, "Brazil," 3.

62. Ibid., 1.

63. Alicia Fazzano and Marc A. Weiss, "Curitiba, Brazil: GUD Metropolitan Economic Strategy Report," *Global Urban Development* (2004): 14.

64. Cassio Taniguchi, "Transport and Urban Planning in Curitiba," *disP-The Planning Review* 37, no. 147 (2001): 14–19.

65. J. A. Ribeiro and N. Tavares, *Curitiba: A revolução ecológica*, trans. N. Torres (Curitiba: Prefeitura Municipal de Curitiba, 1992).

66. International Council for Local Environmental Initiatives, "Curitiba: Orienting Urban Planning to Sustainability," Municipality of Curitiba, Brazil, Case Study 77, May 2002, 4, http://www .iclei.org.br/polics/CD/P2_4_Estudos%20de%20Caso/1_Planejamento%20Urbano/PDF106_EC77 _Curitiba_ing.PDF.

67. Trust for Public Land, "2011 City Park Facts," 2011, http://cloud.tpl.org/pubs/ccpe-city-park -facts-2011.pdf.

68. Gustafsson and Kelly, "Urban Innovations in Curitiba," 18.

69. "Success Stories: Sustainable Urban Planning in Brazil," United Nations Environment Programme, Green Economy Institute, http://web.unep.org/greeneconomy/success-stories-8.

70. Imam Anshori, Chief of Sub Directorate of River Basin Management, Directorate General of Water Resources, Ministry of Public Works, "Basin Water Resources Management and Organization in Indonesia, ca. 2005," http://www.narbo.jp/data/01_events/materials/tc02_2_06b.pdf.

71. Asian Development Bank, "Country Water Action: Indonesia Simple Solution for Drinking Water Makes Big Difference," March 2006, https://www.scribd.com/doc/110969046/Indonesia -Water-Action-Simple-Solution-for-Drinking-Water-Makes-Big-Difference.

72. Water and Sanitation Program, The World Bank, "Water Supply and Sanitation in Indonesia: Turning Finance into Service for the Future," May 2015, https://www.wsp.org/sites/wsp.org/files /publications/WSP-Indonesia-WSS-Turning-Finance-into-Service-for-the-Future.pdf.

73. Asian Development Bank, "Country Water Action."

74. ICLEI Japan and Japan Fund for Global Environment, "Bogor, Indonesia: Using Local Initiative and Knowledge for Local Water and Soil Conservation," Case Study 129, April 2011, http:// www.iclei.org/fileadmin/PUBLICATIONS/Case Studies/ICLEI_cs_129_Bogor_2011.pdf.

75. Ibid.

76. Ibid.

77. Devisanthi Tunas, "The 'Biopori' Phenomenon: Self-Reliance in Indonesian Capital in the Face of Annual Flooding," *Green Asia Force*, February 28, 2013, http://www.greenasiaforce .com/Blog/the-biopori-phenomenon-self-reliance-in-indonesian-capital-in-the-face-of-annual -flooding/.

78. Kris Tjokro, "Probolinggo Residents Create Biopori Holes," *TNOL Asia*, September 2, 2014, http://www.tnol.asia/environment/20504-probolinggo-residents-create-biopori-holes.html.

79. Nugroho Nurdikiawan, "Indonesia: Communities Work to Improve Sanitation and Clean Water," *World Bank News*, March 22, 2014, http://www.worldbank.org/en/news/feature/2014/03 /21/indonesia-communities-work-to-improve-sanitation-access-to-clean-water.

80. Nadya Natahadibrata, "All Indonesia to Have Access to Clean Water by 2019," *Jakarta Post*, Jakarta Business, May 14, 2014, http://www.thejakartapost.com/news/2014/05/14/all-indonesians -have-access-clean-water-2019.html#sthash.LFuPkDi5.dpuf.

81. World Bank News, "Ethiopia Climate Project Receives Africa's First Forestry Carbon Cred- its under the CDM," October 9, 2012, http://www.worldbank.org/en/news/feature/2012/10/09 /ethiopia-climate-project-receives-africa-s-first-forestry-carbon-credits.

82. Africa News Agency, "Ethiopian Project Sets World Climate Change Example," *afrol News*, March 4, 2012, http://www.afrol.com/articles/35532.

83. Byamukama Biryahwaho and Michael Misiko, with Hailu Tefera and Assefa Tofu, "Insti- tutional Innovations in African Smallholder Carbon Projects. Case Study: Humbo Ethiopia Assisted Natural Regeneration Project," Climate Change, Agriculture and Food Security Research Program, June 2012, http://r4d.dfid.gov.uk/PDF/Outputs/CCAFS/AfricanAgCarbon-CaseStudy -Humbo.pdf.

84. Ibid.

85. Africa News Agency, "Ethiopian Project."

86. Ibid.

87. Project Design Document Form for Afforestation and Reforestation Project Activities (CDM- AR-PDD) Version 04, 1–102 at 14; Clean Development Mechanism, United Nations Framework Convention on Climate Change, http://www.netinform.net/KE/files/pdf/PDD_AR_Plantar.pdf.

88. World Bank News, "Ethiopia Climate Project."

89. Carolyn Kaboré, "Evaluation of a Community Managed Forest Project: Humbo, Ethio- pia," Better Evaluation Organization, July 2013, http://betterevaluation.org/sites/default/files /Evaluation%20of%20a%20community%20managed%20forest%20project%20-%20Caro lyn%20Kabore_0.pdf.

90. World Resources Institute, "African Countries Launch AFR100 to Restore 100 Million Hectares of Land," Press Release, December 2015, http://www.wri.org/news/2015/12/release-african-countries-launch-afr100-restore-100-million-hectares-land.

91. The failures may also be attributable to the preference for centralized, one-time investments, which may flow from a desire to emulate the systems of industrialized societies.

92. Simon Toze, "Reuse of Effluent Water-Benefits and Risks," *Agricultural Water Management* 80 (2006): 147–159 at 148.

93. Dhrubajyoti Ghosh, "Wastewater-Fed Aquaculture in the Wetlands of Calcutta—An Overview," *Proceedings of the International Seminar on Wastewater Reclamation and Reuse for Aquaculture*, Calcutta, India, December 6–9, 1988, http://www.bvsde.ops-oms.org/bvsair/e/repindex/repi53/calcutta/calcutta.html.

94. Nitai Kundu, Mausumi Pal, and Sharmistha Saha, "East Kolkata Wetlands: A Resource Recovery System through Productive Activities," *Proceedings of Taal 2007: The 12th World Lake Conference*, 2008, 868–881.

95. Ghosh, "Wastewater-Fed Aquaculture," 6.

96. Guatam Gupta and Vivekananda Mukherjee, "Valuation of an Ecosystem and Its Impact on Livelihood Support: The Case Study of East Calcutta Wetlands," Third Progress Report, October 15, 2009, South Asian Network for Development and Environmental Economics (SANDEE), SANDEE online, http://www.sandeeonline.com/uploads/forums/21_FRM1_Third_Report.pdf.

97. S. W. Bunting, "Confronting the Realities of Wastewater Aquaculture in Peri-urban Kolkata with Bioeconomic Modelling," *Water Research* 41 (October 2006): 499–505.

98. Amanda Suutri, "India—East Calcutta—Making the Most of It: Wastewater, Fishponds, and Agriculture," The EcoTipping Points Project, July 2006, http://ecotippingpoints.org/our-stories/indepth/india-calcutta-wetland-wastewater-agriculture-fishpond.html.

99. Vivekananda Mukerjee and Gautum Gupta, "Cultivation under Wastewater Irrigation: The Case of the East Calcutta Wetlands," South Asian Network for Development and Environmental Economics Working Paper no. 62–11, August 2011, 15.

100. Santosh Ghosh, "Indigenous and Ecological Development of Wetlands in the Kolkata Metropolitan Area," *Proceedings of the 2006 River Symposium*, http://archive.riversymposium.com/2006/index.php?element=06GHOSHSantoshKumar.

101. Kolkata Environmental Improvement Project (KEIP), "The East Kolkata Wetland System," http://www.keip.in/bl333/wetlands.php.

102. R. Sarkar, "Valuing the Ecosystem Benefits of Treatment of Manmade Wetlands Using Conventional Economic Indicators: A Case Study of the East Calcutta Wetlands," Occasional Papers no. 1, 2002, Department of Business Management, University of Calcutta.

103. Anindita Bhattacharyya, Subro Sen, Pankj Kumar Roy, and Aziz Mazumdar, "A Critical Study on Status of East Kolkata Wetlands with Special Emphasis on Water Birds as Bio-indicator,"

Proceedings of Taal 2007: The 12th World Lake Conference, 2008, 1561–1570, http://www.moef.nic .in/sites/default/files/nlcp/P%20-%20World%20Case%20Studies/P-31.pdf.

104. Ibid., 1561.

105. Ghosh, "Wastewater-Fed Aquaculture."

106. U.S. Environmental Protection Agency, "The US Government's Global Methane Initiative Accomplishments Report 2013," 21, https://www.epa.gov/sites/production/files/2016-01/docu ments/usg_2013_accomplishments_0.pdf.

107. International Energy Agency, IEA Bioenergy Task 37 Country Reports of Member Coun- tries, Istanbul, April 2011, http://www.ieabcc.nl/publications/IEA%20Bioenergy%202011%20 Annual%20Report.pdf.

108. H. H. P. Fang and Y. Liu, "Anaerobic Wastewater Treatment in (Sub-) Tropical Regions," in *Advances in Water and Wastewater Treatment Technology: Molecular Technology, Nutrient Removal, Sludge Reduction and Environmental Health*, ed. T. Matsuo (Amsterdam: Elsevier Science, 2001), 285.

109. Joseph Frijns, Jan Hofman, and Maarten Nedrlofl, "The Potential of (Waste)water as Energy Carrier," *Energy Conversion and Management* 65 (2013): 357–363.

110. Wim J. van Nes. "Biogas from Anaerobic Digestion Rolls Out across Asia," *Renewable Energy World* (January–February 2006): 102–111.

111. Christian Mueller, "Decentralised Co-digestion of Faecal Sludge with Organic Solid Waste," Sandec, Department of Water and Sanitation in Developing Countries at the Swiss Federal Insti- tute for Aquatic Science and Technology, 2009, 1–76, http://www.eawag.ch/forschung/sandec /publikationen/swm/dl/Mueller_2009.pdf.

112. Ibid. Those biodigesters that include animal dung generate more biogas than those taking sewage only.

113. Ibid. In the periurban areas around Maseru, yard waste and other organic solids are fed into the digesters, which reduces urban litter and improves overall sanitation.

114. Seetharam C. Jansi and Santosh K. Mishra, "Wastewater Treatment and Reuse: Sustainability Options," *Consilience: The Journal of Sustainable Development* 10, no. 1 (2013): 1–15.

115. Mueller, "Decentralised Co-digestion of Faecal Sludge."

116. Van Nes, "Biogas," 104.

117. U.S. Government Global Methane Initiative, "Wastewater Success Story: La Farana Wastewa- ter Plant Biogas Upgrade," https://www.globalmethane.org/expo-docs/posters/Wastewater/WW _CL_Success_La-Farfana-WWTP_FINAL.pdf.

118. C. J. Schuster-Wallace, C. Wild, and C. Metcalfe, "Valuing Human Waste as an Energy Resource: A Research Brief Assessing the Global Wealth in Waste," United Nations University

Institute for Water, Environment and Health (UNU-INWEH), 2015, 1–14 at 8, http://inweh.unu.edu.

119. Gleick, "Global Freshwater Resources."

120. Stockholm Environment Institute, "Rainwater Harvesting," 5.

Chapter 4

1. Vaclav Smil, "World History and Energy," in *Encyclopedia of Energy*, vol. 6 (Cambridge, MA: Elsevier Inc., 2004), 549.

2. Carbon Dioxide Information Analysis Center, "Gapminder World V0.0.7," 2012.

3. Ibid.

4. Greenhouse gas emissions are conventionally measured in metric tons.

5. The World Bank, *World Development Indicators* (Washington, DC: The World Bank, 2012).

6. Ibid.

7. There are of course scenarios in which the level of service is too high (e.g., air conditioners set to 65°F), but waste and overconsumption tend not to be predominant problems in emerging economies.

8. The Housing Development Agency, "South Africa: Informal Settlements Status" (Johannesburg: Housing Development Agency, 2012), 7–9.

9. Emma Bryce, "South African Settlement Welcomes the iShack," *New York Times*, November 5, 2012, http://green.blogs.nytimes.com/2012/11/05/south-african-settlement-welcomes-the-ishack/.

10. Camile Bautista, "iShack Brings Sustainable Living to South African Slums," *Mashable*, January 9, 2013, http://mashable.com/2013/01/09/ishack-sustainable-living-south-africa/.

11. Teo Kermeliotis, "High-Tech Shack Brings Solar Power to Slums," *CNN*, January 8, 2013, http://www.cnn.com/2013/01/08/tech/innovation/ishack-slums-south-africa/.

12. iShack project, http://www.ishackproject.co.za/.

13. The World Bank, "Micro-hydro Lights Way for Villages," http://go.worldbank.org/WAGCQC0H22.

14. K.C Surendra, Samir Kumar Khanal, Prchand Shrestha, Buddhi Lamsal, "Current Status of Renewable Energy in Nepal: Opportunities and Challenges," *Renewable and Sustainable Energy Reviews* 15 (2011): 4107–4117 at 4107.

15. Amy Yee, "Microhydro Drives Change in Rural Nepal," *New York Times*, June 20, 2012.

16. Government of Nepal, *Rural Energy Policy 2006* (Kathmandu: Ministry of Environment, 2006).

17. Deepak Adhikari, "Hydropower Development in Nepal," *Nepal Rastra Bank Economic Review* 18 (2006): 84.

18. Ibid.

19. The benefits of self-reliance are discussed further in chapter 6, which focuses on participatory development.

20. United Nations Development Programme, *Scaling Up Decentralized Energy Services in Nepal* (New York: United Nations, 2011), 6.

21. Udit Misra, "Omnigrid's Unique Solar Power Model," *Forbes India*, May 23, 2013, http://forbesindia.com/article/work-in-progress/omnigrids-unique-solar-power-model/35267/1.

22. World Energy Outlook, WEO Database, 2015, http://www.worldenergyoutlook.org/resources/energydevelopment/energyaccessdatabase/.

23. Adam Wills, "OMC Power," Mobile for Development (M4D), 2012, http://www.m4dimpact.com/analysis/case-studies/omc-power.

24. Coco Liu, "Renewable Energy, Once a Dream, Lights Up Some of India's Slums," *ClimateWire*, December 20, 2013, http://www.eenews.net/stories/1059992166.

25. R. Srinivasan, "In Areas Where Grid Power and Diesel Supply Is [*sic*] Erratic, Solar Power Has a Strong Business Case," *Power Watch India*, June 4, 2015, http://powerwatchindia.com/in-areas-where-grid-power-and-diesel-supply-is-erratic-solar-power-has-a-strong-business-case/.

26. Katherine Tweed, "Why Cellular Towers in Developing Nations Are Making the Move to Solar Power," *Scientific American*, January 15, 2013, http://www.scientificamerican.com/article/cellular-towers-moving-to-solar-power/.

27. Telecom Regulatory Authority of India, "Highlights of Telecom Subscription Data as of 28th February, 2015," New Delhi, April 10, 2015, http://www.trai.gov.in/WriteReadData/PressRealease/Document/PR-TSD-Feb-10042015.pdf.

28. Justin Guay, "India: Forget the Grid, Community Power Is Here," *Compass: Pointing the Way to a Clean Energy Future*, Sierra Club, August 2012, http://sierraclub.typepad.com/compass/2012/08/india-forget-the-grid-community-power-is-here.html#more.

29. Justin Guay, "Meet Uttar Pradesh, India's Next Distributed Solar Hotbed," *Sierra Club, India Environment Post*, November 19, 2013, http://action.sierraclub.org/site/MessageViewer?em_id=265265.0&dlv_id=224822. Typically, the plants take up 5,000 to 10,000 ft^2 of rented land, but they could be scaled up. See Misra, "Omnigrid's Unique Solar Power Model."

30. OMC's most recently constructed plants are designed to incorporate wind or biogas energy as well and have battery banks with diesel-fired backup generators. See Pär Almqvist, "Cutting the Cord for a Better Energy Future," *The Solutions Journal* 4, no. 2 (June 2013), http://www.thesolutionsjournal.org/node/22511?page=31%2C0%2C0%2nC0%2C0%2C1&quicktabs_1=0.

31. Wills, "OMC Power."

32. Guay, "Meet Uttar Pradesh."

33. Wills, "OMC Power."

34. Misra, "Omnigrid's Unique Solar Power Model."

35. Guay, "India."

36. Misra, "Omnigrid's Unique Solar Power Model."

37. Ibid.

38. Ibid.

39. Liu, "Renewable Energy."

40. G. Ardizzon, G. Vavazzin, and G. Pavesi, "A New Generation of Small Hydro and Pumped-Hydro Power Plants: Advances and Future Challenges," *Renewable and Sustainable Energy Reviews* 31 (2014): 749.

41. "Energy Storage: Packing Some Power," Technological Quarterly, *The Economist*, March 3, 2011, http://www.economist.com/node/21548495?frsc=dg%7Ca.

42. The lower reservoir has a capacity of 150,000 m^3 (39.62 million gal), and the upper reservoir has a capacity of 380,000 m^3 (132 million gal). See R. Godina, E. M. G. Rodrigues, J. C. O. Matias, and J. P. S. Catalão, "Sustainable Energy System of El Hierro Island," presented at the International Conference on Renewable Energies and Power Quality, May 25–27, 2015, *Renewable Energy and Power Quality Journal* no. 13 (April 2015): 2.

43. SMILEGOV "El Hierro Project, a Successful Multilevel Governance Good Practice Resulting in a 100% Renewable Energy-Isolated Territory," Smart Multilevel Governance: Island News, http://www.sustainableislands.eu/news/islandnews/hierro-spanish-canarias-multilevel-governance-good-practice-coopernergy.html.

44. Godina et al., "Sustainable Energy System," 5.

45. Laurie Guevara-Stone, "How a Small Spanish Island Became a Renewable Energy Pioneer," *GreenBiz*, March 3, 2014, https://www.greenbiz.com/blog/2014/03/03/how-small-spanish-island-became-renewable-energy-pioneer-el-hierro.

46. Sonal Patel, "A Spanish Island's 100% Wind-and-Water Power Solution," *Power*, August 1, 2014, http://www.powermag.com/a-spanish-islands-100-wind-and-water-power-solution/.

47. Guevera-Stone, "Small Spanish Island."

48. Deutsche Welle, "Spanish Island Generates Its Own Clean Energy," Newsletter, *Top Stories—Environment*, November 18, 2014, http://www.dw.com/spanish-island-generates-its-own-clean-energy/a-18032899.

49. IEEE, "A Wind–Hydro–Pumped Storage Station Leading to High RES Penetration in the Autonomous Island System of Ikaria," http://ieeexplore.ieee.org/xpl/freeabs_all.jsp?arnumber=5512672.

50. Tim Forcey and Dylan McConnell, "Pumped Hydro—The Forgotten Storage Solution," *RenewEconomy*, July 2, 2014, http://reneweconomy.com.au/2014/pumped-hydro-the-forgotten -storage-solution-47248.

51. PennEnergy, "Developer Files Storage, Solar Power Project Application with FERC," *PennEnergy Newsletter*, February 27, 2014, http://www.pennenergy.com/articles/pe/2014/02/developer-files -storage-solar-power-project-application-with-ferc.html.

52. Matt Lucky, "Get Pumped (Hydro) for Wind Energy in Haiti," September 16, 2011, World-watch blog, Worldwatch Organization, http://blogs.worldwatch.org/revolt/get-pumped-for-wind -energy-in-haiti/.

53. Water.org, http://water.org/country/haiti/.

54. Lucky, "Get Pumped."

55. Lester R. Brown, "Stabilizing Climate: Shifting to Renewable Energy," in *Plan B 4.0: Mobilizing to Save Civilization* (New York: Earth Policy Institute and W. W. Norton, 2009), 109–142 at 122.

56. Anselm Eisentraut and Adam Brown, "Heating without Global Warming: Market Devel-opments and Policy Considerations for Renewable Heat," *International Energy Agency* 19 (2014): 1–92, https://www.iea.org/publications/freepublications/publication/FeaturedInsight_Heating WithoutGlobalWarming_FINAL.pdf.

57. Peter Newborne and Bryn Welham, "Joining the Grid: Sustainable Energy in Brazil," *Over-seas Development Institute*, August 2014, 23, https://www.odi.org/sites/odi.org.uk/files/odi-assets /publications-opinion-files/9149.pdf.

58. U.S. Energy Information Administration, "Brazil Overview," https://www.eia.gov/beta /international/analysis_includes/countries_long/Brazil/brazil.pdf.

59. Ibid., 11.

60. Helena F. Naspolini, H. S. G. Militão, and R. Rüther, "The Role and Benefits of Solar Water Heating in the Energy Demands of Low-Income Dwellings in Brazil," *Energy Conversion and Man-agement* 51, no. 12 (2010): 2835–2925.

61. ICLEI Case Studies, Betim, Brazil, "Solar Heaters in Low Income Housing: Energy and Finan-cial Savings," 2, http://www.iclei.org/fileadmin/PUBLICATIONS/Case_Studies/ICLEI_cs_112_Betim _ENG_August_2010.pdf.

62. Ibid., 3.

63. Ibid.

64. Geraldo Antunes, "Reference Center in Renewable Energy: Renewable Energy in Betim," pre-sentation for the International Conference on Local Renewables, April 27–29, 2009, Freiburg im Breisgau, Germany, 35, http://series.local-renewables-conference.org/fileadmin/template/events/lr _freiburg_2009/files/Presentations/Antunes_P4.pdf.

65. Renewable Energy Policy Network for the 21st Century (REN21), *Renewables 2010 Global Status Report* (Paris: REN21 Secretariat, 2010), 41.

66. Antunes, "Reference Center," 21–26.

67. REN21, *Renewables 2010 Global Status Report*, 41.

68. Andrew Orkin, Joanna Birenbaum, and the Grand Council of the Crees, "Pushed to the Edge of Extinction: Racism against Indigenous Peoples in Canada," Grand Council of the Crees, 2001, http://www.gcc.ca/pdf/INT000000005.pdf.

69. Oujé-Bougoumou, "Our Innovation," http://www.ouje.ca/innovation.

70. Ibid.

71. Ibid.

72. Biomass Energy Resource Center, "Waste Wood Helps Fuel a Native Community's Development," BERC Case Study Series, http://www.biomasscenter.org/images/stories/ouje-bougoumou.pdf.

73. Ministry of Natural Resources, "RETScreen International Clean Energy Decision Support Center," http://ww.retscreen.net/download.php/da/79/3/BIOH03-C.pdf.

74. Ibid.

75. Ibid.

76. Biomass Energy Resource Center, "Waste Wood."

77. This approach has been demonstrably effective. In Nepal, for example, after biogas systems were implemented, the use of wood fuel dropped by 53 percent. See H. Katuwal and A. K. Bohara, "Biogas: A Promising Renewable Technology and Its Impact on Rural Households in Nepal," *Renewable and Sustainable Energy Reviews* 13, no. 9 (2009): 2668–2674. doi:10.1016/j.rser.2009.05.002.

78. Ibid.

79. Yvonne Vögeli, Christian Riu Lohri, Amalia Gallardo, Stefan Diener, and Christian Zurbrügg, *Anaerobic Digestion of Biowaste in Development Countries: Practical Information and Case Studies* (Dübendorf: Eawag—Swiss Federal Institute of Aquatic Science and Technology, 2014), 8.

80. For biodigestion of human waste and animal manure, temperature and retention time determine whether the effluent can be used in food production without health risk.

81. Vögeli et al., *Anaerobic Digestion*, 8.

82. Lennart de Groot and Anne Bogdanski, *Bioslurry=Brown Gold?: A Review of Scientific Literature on the Co-product of Biogas Production* (Rome: Food and Agriculture Organization of the United Nations, 2003), iv.

83. C2ES—Center for Climate and Energy Solutions, http://www.c2es.org/technology/factsheet/anaerobic-digesters; Siyenza Management for Africa Energy Indaba, "Turning Waste into Energy: The

Win-Win-Win Solution," *ENERGY Digital—Renewables*, January 16, 2012, http://www.energydigital .com/renewables/3093/Turning-Waste-into-Energy:-the-WinWinWin-solution.

84. Fredrick Ochieng, "Enhanced Energy Access for the Urban Poor: Best Practice Casebook," United Nations Habitat, 30, http://mirror.unhabitat.org/downloads/docs/Enhanced%20Energy %20Access%20for%20Urban%20poor%20Best%20Practice%20Case%20book.pdf.

85. Valerie J. Brown, "Biogas: A Bright Idea for Africa," *Environmental Health Perspectives* 114, no. 5 (May 2006): A300–A303; United Nations Framework Convention on Climate Change, "Cows to Kilowatts: Anaerobic Bio-digestion of Abattoir Waste Generates Zero Emission and Creates Sustainable Bio-energy and Bio-fertiliser in Africa," http://unfccc.int/secretariat/momentum_for_change /items/7140.php.

86. Julia Steets, "From Cows to Kilowatts—a Case Study in Successful Technology Transfer," *WIPO: World Intellectual Property Organization Magazine*, March 2009, http://www.wipo.int/wipo _magazine/en/2008/02/article_0002.html.

87. Brown, "Biogas."

88. United Nations Sustainable Development Knowledge Platform, http://www.un.org/esa /sustdev/publications/africa_casestudies/kilowatts.pdf.

89. United Nations Framework Convention on Climate Change, "Cows to Kilowatts."

90. Brown, "Biogas."

91. A. Robinson, G. Sewell, N. Damdaran, E. David, and N. Kalas-Adams, "Landfills in Developing Countries and Global Warming," in *Proceedings Sardinia 2003, Ninth International Waste Management and Landfill Symposium*, Cagliari, Italy, CISA, Environmental Sanitary Engineering Centre, October 6–10, 2003, 2.

92. Ibid., 3.

93. World Wildlife Fund, "Belo Horizonte Sustainability," http://wwf.panda.org/what_we_do /footprint/cities/urban_solutions/themes/governance/?228952.

94. Ibid.

95. Ibid.

96. Ibid.

97. Robinson et al., "Landfills in Developing Countries," 2.

98. Gregory T. Pope, "The Seven Wonders of the Modern World," *Popular Mechanics* 172, no. 12 (December 1995): 48–56.

99. Bogumil Terminski, "Development-Induced Displacement and Resettlement: Theoretical Frames and Current Challenges" (Geneva: University of Geneva, 2013), 16.

100. Nelton M. Friedrich, Jorge H. H. El Khouri, and Gustavo O. Rojas, "ITAIPU Binacional: Electrical Integration, Regional Technological Development and Sustainability," presented at

Cooperation for Water, Energy and Food Security in Transboundary Basins under a Changing Climate, Ho Chi Minh, Viet Nam, April 2, 2014, http://www.mrcsummit.org/presentations/track3/3 .3-d-water-energy-and-food-systemic-approach-NeltonMiguel.pdf.

101. Melissa C. P. P. Mathias and João F. C. Mathias, "Biogas in Brazil: A Governmental Agenda," *Journal of Energy and Power Engineering* 9 (2015): 1.

102. Peter Meisen, *Renewable Energy Potential of Brazil* (San Diego, CA: Global Energy Network Institute, 2010), http://www.geni.org/globalenergy/research/renewable-energy-potential-of-brazil /re-pot-of-brazil.pdf.

103. Mathias and Mathias, "Biogas in Brazil," 3.

104. IEA Bioenergy, "Bio-energy in Family Farming: A New Sustainable Perspective for the Rural Sector in Brazil," IEA Energy Technology Network, September 2013, http://www.iea-biogas.net /case-studies.html.

105. Parque Tecnologico Itaipu and Itaipu Binacional, "2012 PTI Results Report," http://www.pti .org.br/sites/default/files/relatorio_resultados_ing.pdf.

106. Carlos H. Coimbra-Araújo, Leidiane Mariane, Cicero Bley Júnior, Elisandro P. Frigo, Michelle S. Frigo, Izabela R. C. Araújo, and Helton J. Alves, "Brazilian Case Study for Biogas Energy: Production of Electric Power, Heat and Automotive Energy in Condominiums of Agro-energy," *Renewable and Sustainable Energy Reviews* 40 (2014): 826–839 at 829.

107. Mario Osava, "Brazilian Hydroelectricity Giant Promotes Biogas," *Terramerica: Environment and Development*, September 2013, 2, http://www.ipsnews.net/2013/09/brazilian-hydroelectricity -giant-promotes-biogas/.

108. United Nations Department of Economic and Social Affairs (UNDESA), "Water for Life," UN-Water Best Practices Award Winners," http://www.un.org/waterforlifedecade/winners2015.shtml.

109. ITAIPU Binacional Cultivating Good Water Program, "Overview: The Itapu Binacional Social-Environmental Program," February 2010, 8, http://www.cultivandoaguaboa.com.br/sites /default/files/iniciativa/CAB_english.pdf.

110. Angela Behrendt, "Energized by Change: A Case Study in Corporate Culture Change at Brazil's Itaipu STP," paper presented at 30th IASP World Conference on Science and Technology Parks, 2013, 4, http://inhalt.com.br/portodigital/Workshop4speaker3Angela_ARG.pdf.

111. Parque Tecnologico Itaipu and Itaipu Binacional, "2012 PTI Results Report."

112. Itaipu Binacional, "Biogas Production Model from Paraná Will Be Replicated in Uruguay," Press Release, August 27, 2013, https://www.itaipu.gov.br/en/press-office/news/biogas-production -model-parana-will-be-replicated-uruguay.

113. Ivan B. T. Lima, Fernando M. Ramos, Luis A. W. Bambace, and Reinaldo R. Rosa, "Methane Emissions from Large Dams as Renewable Energy Resources: A Developing Nation Perspective," *Mitigation and Adaptation Strategies for Global Change* 13, no. 2 (2008): 193–206.

114. Drew T. Shindell, Greg Faluvegi, Dorothy M. Koch, Gavin A. Schmidt, Nadine Unger, and Susanne E. Bauer, "Improved Attribution of Climate Forcing to Emissions," *Science*, no. 326 (October 30, 2009): 717–718.

115. F. M. Ramos, L. W. W. Bambace, I. B. T. Lima, R. R. Rosa, E. A. Mazzi, and P. M. Fernside, "Methane Stocks in Tropical Hydropower Reservoirs as a Potential Energy Source," *Climatic Change* 93 (2009): 1–13 at 2.

116. Ibid., 3.

117. L. A. W. Bambace, F. M. Ramos, I. B. T. Lima, and R. R. Rosa, "Mitigation and Recovery of Methane Emissions from Tropical Hydroelectric Dams," Third Dubrovnik Conference on Sustainable Development of Energy, Water and Environment Systems, in *Energy* 32, no. 6 (June 2007): 1038–1046 at 1040.

118. Paul Brown, "Tropical Dams an Underestimated Methane Source," Climate Central, Climate New Network, September 14, 2014, http://www.climatecentral.org/news/tropical-dams-methane -18019.

119. Ramos et al., "Methane Stocks," 5.

120. Lima et al., "Methane Emissions," 196.

121. Erica Gies, "A Dam Revival, Despite Risks: Private Funding Brings a Boom in Hydropower, with High Costs," *New York Times*, November 19, 2014.

122. Kenneth Agutamba, "Rwanda: Kuvuwatt Energy Finally Connected to National Grid," *New Times* (Kigali), November 17, 2015, http://allafrica.com/stories/201511170533.html.

123. Jonathan W. Rosen, "Lake Kivu's Great Gas Gamble," *Technology Review*, April 16, 2015, https://www.technologyreview.com/s/536656/lake-kivus-great-gas-gamble/.

124. One worthy of mention (and potential replication) is the use of treated biogas for municipal buses in Lille, France, where the production of natural gas from sewage plant methane was coupled with the production of biogas in a municipal organic-waste handling plant (which ultimately wound up hosting the city bus depot, where the buses could refuel right at their own garage). The program owes its success to a number of factors, including political and technical support from the European Union. See Hillary Brown, *Next Generation Infrastructure: Principles for Post-industrial Public Works* (Washington, DC: Island Press, 2014), 60–63.

125. Viet Hung, "Presentation: Danang CDS Addressing Climate Change within Urban Planning Systems in Vietnam," *Proceedings of the Resilient Cities Congress*, ICLEI, 2013, http://resilient -cities.iclei.org/fileadmin/sites/resilient-cities/files/Resilient_Cities_2013/RC2013_Hung_03 .pdf.

126. Institute for Social and Environmental Transition (ISET), "Da Nang, Vietnam Policy Brief, Alternative Development Pathways: Examining the Da Nang Master Plan," 2013.

127. "Soleckshaw, the New Vehicle in Kolkata," *Times of India*, November 6, 2011.

128. Anisha Sinha, "Soleckshaw—Solar Power Motorised Rickshaw," *Next Big What*, April 22, 2010, https://www.nextbigwhat.com/soleckshaw-solar-power-motorised-rickshaw-innovation-297/.

129. Ibid.

130. Narayanan Suresh and Vantika Dixit, "Solar Powered Cycle," *MIT Technology Review*, July 2009, http://m.technologyreview.com/computing/23043/.

131. ICLEI Publications, "Volta Redonda: Cooking Oil Waste Is Used in Community Development with the Eco-oil Program," ICLEI Case Study 107, 2009, http://www.iclei.org/fileadmin/PUBLICATIONS/Case_Studies/ICLEI_cs_106_Volta_Redonda_2009.pdf.

132. Jeb Brugmann, "Operationalizing the Urban NEXUS: Towards Resource-efficient and Integrated Cities and Metropolitan Regions," Case Story Series (Bonn: Deutsche Gesellschaft für Internationale Zusammenarbeit [GIZ] GmbH, 2014), 1–70 at 26.

133. Ibid.

134. Jonathan Watts, "Uruguay Makes Dramatic Shift to Nearly 95% Electricity from Clean Energy," *The Guardian*, December 3, 2015, https://www.theguardian.com/environment/2015/dec/03/uruguay-makes-dramatic-shift-to-nearly-95-clean-energy.

135. Joe Thwaites, "How Uruguay Became a Wind Power Powerhouse," *Renewable Energy World*, March 16, 2016, http://www.renewableenergyworld.com/articles/2016/03/how-uruguay-became-a-wind-power-powerhouse.html.

136. Waste-to-energy technologies may be considered a longer-term solution. Energy-from-waste (EFW) has been successfully adopted across much of the European Union, as well as in Japan, China, and South Korea. EFW plants could be encouraged, if not incentivized by banks and lending partners, in those developing regions with a balanced municipal solid-waste mix. However, in many regions, typically high percentages of wet organic waste in the urban mix make this technology more expensive and less reliable. This will inevitably change with improved living standards and goods consumption as other types of materials are introduced into the waste stream.

137. Andreas Goldthau, "Rethinking the Governance of Energy Infrastructure: Scale, Decentralization and Polycentrism," *Energy Research and Social Science* 1 (2014): 134–140.

138. Dilip Ahuja and Marika Tatsutani, "Sustainable Energy for Developing Countries, SAPIENS," *Surveys and Perspectives Integrating Environment and Society* 2, no. 1 (2009): 1–16 at 9.

139. Bloomberg New Energy Finance Group with the Multilateral Investment Fund Part of the Inter-American Development Bank, "Climatescope 2014 Report: Mapping the Global Frontiers for Clean Energy Investment," 1–211 at 5.

Chapter 5

1. The World Bank, *World Development Report: Development and Climate Change* (Washington, DC: World Bank, 2010).

2. Ibid., 13.

3. I. Burton, E. Diringer, and J. Smith, "Adaptation to Climate Change: International Policy Options," Pew Center on Global Climate Change, 2006, 1–28, http://www.c2es.org/docUploads /PEW_Adaptation.pdf.

4. Barry Smit and Olga Pilifosova, "Adaptation to Climate Change in the Context of Sustainable Development and Equity," *Sustainable Development* 8, no. 9 (2003): 879–912 at 879.

5. Justin Lifu Lin, "Keynote Address: Development and Climate Change," in *Climate Governance and Development*, ed. Albrecht Ansohn and Boris Pleskovic (Washington DC: The World Bank, 2011), 1–170 at 28.

6. The World Bank, *The Cost to Developing Countries of Adapting to Climate Change: New Methods and Estimates* (Washington, DC: World Bank, 2010), 1.

7. Ibid., 33.

8. Damien Evans, Christophe Pottier, Roland Fletcher, Scott Hensley, Ian Tapley, Anthony Milne, and Michael Barbetti, "A Comprehensive Archaeological Map of the World's Largest Preindustrial Settlement Complex at Angkor, Cambodia," *Proceedings of the National Academy of Sciences* 104, no. 36 (2007): 14277–14282.

9. Roland Fletcher, Damien Evans, Ian Tapley, and Anthony Milne, "Angkor: Extent, Settlement and Ecology: Preliminary Results of an AIRSAR Survey," *Indo-Pacific Prehistory Association Bulletin Taipei Papers* 2 (2000): 137.

10. Roland Fletcher, Christophe Pottier, Damian Evans, and Matti Kummu, "The Development of the Water Management System of Angkor: A Provisional Model," *Bulletin of the Indo-Pacific Prehistory Association* 28 (2008): 57–66.

11. Richard Stone, "Divining Angkor: Cultural Heritage," *National Geographic* 216, no. 1 (July 2009), 4.

12. Damien et al., "Comprehensive Archaeological Map," 14280.

13. Elfren B. Paz, "Learning from the Experiences of Past Cultures, Lost and Abandoned Urban Centres," *Assumption University Journal of Technology* 8, no. 3 (2005): 136.

14. Karen J. Coates, *Cambodia Now: Life in the Wake of War* (Jefferson, NC: McFarland, 2005), 51.

15. Paz, "Lost and Abandoned Urban Centers," 135–145.

16. Roland Fletcher, "Low-Density, Agrarian-Based Urbanism: A Comparative View," *Insights, University of Durham* 2, no. 4 (2009): 2–19.

17. Brendan M. Buckley, Kevin J. Anchukaitis, Daniel Penny, Roland Fletcher, Edward R. Cook, Masaki Sano, Aroonrut Wichienkeeo, Ton That Minh, and Truong Mai Hong, "Climate as a Contributing Factor in the Demise of Angkor, Cambodia," *Proceedings of the National Academy of Sciences* 107, no. 15 (2010): 6748–6752.

18. Udom Hong, "The Angkor Empire: Environment and Conflict," Inventory of Conflict and the Environment Case Study No. 207 (2007), 7, http://www1.american.edu/ted/ice/ankorwat.htm.

19. Terry Lustig, Roland Fletcher, Matti Kummu, Christophe M. Pottier, and Dan Penny, "Did Traditional Cultures Live in Harmony with Nature? Lessons from Angkor, Cambodia," in *Modern Myths of the Mekong*, ed. Matti Kummu, Marko Keskinen, Olli Varis (Helsinki: Helsinki University of Technology, 2008), 81–94.

20. City Climate Leadership Awards, "Jakarta: Socially-Inclusive Coastal Protection Today and for 2030," 2014, http://www.c40.org/profiles/2014-jakarta.

21. Indra M. Surbakti, Izhar C. Idroes, Hendricus A. Simarmata, and Tommy Firman, "Jakarta City Report: Information Related to Climate Change in Jakarta City," paper prepared for the Workshops of Climate Change Vulnerability Assessment and Urban Development Planning for Asian Coastal Cities, Rose Garden, Sampran Riverside, Nakorn, 1–24 at 7, http://startcc.iwlearn.org/doc/Doc_eng_18.pdf.

22. Camilo Mora, Abby G. Frazier, Ryan J. Longman, Rachel S. Dacks, Maya M. Walton, Eric J. Tong, Joseph J. Sanchez, Lauren R. Kaiser, Yuko O. Stender, James M. Anderson, Christine M. Ambrosino, Iria Fernandez-Silva, Louise M. Giuseffi, and Thomas W. Giambelluca, "The Projected Timing of Climate Departure from Recent Variability," *Nature* 502 (October 10, 2013): 183–187, http://www.soc.hawaii.edu/mora/PublicationsCopyRighted/Cities%20Timing.html.

23. Surbakti et al., "Jakarta City Report," 7.

24. Ibid.

25. Ibid., 4.

26. Ibid., 19.

27. Judy L. Baker, *Climate Change, Disaster Risk and the Urban Poor: Cities Building Resilience for a Changing World* (Washington, DC: World Bank Publications, 2012), 1–16, http://elibrary.worldbank.org/doi/abs/10.1596/978-0-8213-8845-7.

28. Bill Tarrant, "The Great Wall of Jakarta," Reuters, December 22, 2015, http://www.reuters.com/article/2014/12/22/us-environment-jakarta-seawall-idUSKBN0K017420141222.

29. Lenny T. Tambun, Novy Lumanauw, and Bayu Marhaenjati, "Jakarta's Flood Problem Totally Solvable, for $9.2 B," *Jakarta Globe*, February 11, 2015, http://jakartaglobe.beritasatu.com/news/jakarta/jakartas-flood-problem-totally-solvable-9-2b/.

30. Urban Innovations, "The Pluit Reservoir Revitalization Project: Socially Inclusive Coastal Protection Today and for 2030," 2014, http://www.urban-innovations.org/index.php/The_Pluit_Reservoir_Revitalization_Project:_(Socially-Inclusive_Coastal_Protection_Today_and_for_2030.

31. Tarrant, "Great Wall of Jakarta."

32. Reliefweb, "India: Orissa Cyclone Appeal No. 28/1999, Final Report International Federation of Red Cross and Red Crescent Societies," November 13, 2002, http://reliefweb.int/report/india/india-orissa-cyclone-appeal-no-281999-final-report.

33. UNEP-WCMC, "In the Front Line: Shoreline Protection and other Ecosystem Services from Mangroves and Coral Reefs," 2006, 1–33, http://www.unep.org/pdf/infrontline_06.pdf.

34. Ibid.

35. UNEP, Maija Bertule, Gareth James Lloyd, and Louise Korsgaard, "Green Infrastructure Guide for Water Management: Ecosystem-Based Management Approaches for Water-Related Infrastructure Projects," *United Nations Environment Programme* 47 (2014): 1–76, http://www.unepdhi .org/-/media/microsite_unepdhi/publications/documents/unep/web-unep-dhigroup-green -infrastructure-guide-en-20140814.pdf.

36. Marcel Marchand, "Mangrove Restoration in Vietnam: Key Considerations and a Practical Guide," *Deltares* 8 (2008): 1–29, http://repository.tudelft.nl/islandora/object/ 98b5ba43-1452-4631-81dc-ad043ef3992c/?collection=research.

37. UNEP-WCMC, "Shoreline Protection."

38. Karen C. Seto and Michail Fragkiasc, "Mangrove Conversion and Aquaculture Development in Vietnam: A Remote Sensing–Based Approach for Evaluating the Ramsar Convention on Wetlands," *Global Environmental Change* 17 (2007): 486–500.

39. Marchand, "Mangrove Restoration in Vietnam," 17.

40. Ibid.

41. International Federation of Red Cross and Red Crescent Societies (IFRC), "Case Study: Mangrove Plantation in Viet Nam: Measuring Impact and Cost Benefit," 1–9 at 3, https://www.ifrc.org /Global/Publications/disasters/reducing_risks/Case-study-Vietnam.pdf.

42. Ibid.

43. Bui Xuan Thong, "Main Characteristics of Significant Wave Formation in the Coastal Zone of Vietnam," 9th International Workshop on Wave Hindcasting and Forecasting, Victoria, BC, Canada, 2006, 1–8.

44. International Federation of Red Cross and Red Crescent Societies, "Mangrove Plantation in Viet Nam."

45. Ibid., 7.

46. Jared O. Bosire, Farid Dahdouh-Guebas, M. Walton, B. I. Crona, R. R. Lewis, C. Field, James G. Kairo, and Nico Koedam, "Functionality of Restored Mangroves: A Review," *Aquatic Botany* 89, no. 2 (2008): 251–259.

47. International Federation of Red Cross and Red Crescent Societies, "Mangrove Plantation in Viet Nam," 3.

48. Ibid., 7.

49. The CARE project is discussed further in chapter 6.

50. Tahera Akter, "Climate Change and Flow of Environmental Displacement in Bangladesh," Unnayan Onneshan—the Innovators, 2009, 1–18 at 10, http://www.unnayan.org/documents /Climatechange/climate_change_flow_environmental_displacement.pdf.

51. Ibid.

52. The World Bank, "Economics of Adaptation to Climate Change: Bangladesh," Washington, DC, 2010, 68, http://documents.worldbank.org/curated/en/841911468331803769/Main-report.

53. Ibid., 73.

54. Ibid.

55. Md. Altaf Hossain, "Indigenous Technology for Adapting to Water Logging Situation for Sustainable Livelihood Security in Low Lying Areas of Bangladesh," *19th World Congress of Soil Science, Social Solution for Changing World* (2010): 122–123.

56. Atiq Rahman, "Floating Vegetable Bed Cultivation," in *Climate of Coastal Cooperation*, ed. Robert Missdorp (Leiden, the Netherlands: Coastal and Marine Union—EUCC, 2011), 179–180, http://www.coastalcooperation.net/part-III/III-3-3-8.pdf.

57. Irin, "Bangladesh: Spreading the Floating Farm's Tradition," Humanitarian News and Analysis, http://www.irinnews.org/report/90002/bangladesh-spreading-the-floating-farms-tradition.

58. Zhaohua Li, Yanqiang Li, Wei Huang, and Lianhai Sang, "Artificial Floating Islands: A Technology for Clean Water and Agricultural Production in Rural China," *International Journal of Environmental Protection* 3, no. 1 (2013): 7–14.

59. Ibid.

60. Haseeb Md. Irfanullah, "Romancing the Floating Garden," *Thomas Reuters Foundation News*, August 29, 2013, http://www.trust.org/item/20130829113321-2x5ao/.

61. Noah Rayman, "This Floating City May Be the Future of Coastal Living," *Time*, June 26, 2014, http://time.com/2926425/the-floating-homes/.

62. O. Adelekan, "Vulnerability of Poor Urban Coastal Communities to Climate Change in Lagos, Nigeria," in *Fifth Urban Research Symposium: Cities and Climate Change: Responding to the Urgent Agenda*, ed. David Satterthwaite and Jane Bicknell (Marseille, France: Human Settlements Programme, International Institute for Environment and Development, 2009), 1–18 at 6.

63. Kunlé Adeyemi and Ore Disu, "Floating City Solutions for Africa's Vulnerable Coastal Communities: The Case of Makoko," in *Perspectives: What Are Sustainable African Cities?* ed. Layla Al-Zubaidi and Jochen Luckschieter (Berlin: Heinrich Böll Foundation, 2012), 18–23 at 21.

64. Ibid., 20.

65. Ibid.

66. Lookman Oshidi, "Makok/Iwaya Waterfront Community: From Risk to Resilience through Regeneration Plan," July 10, 2014, https://oshlookman.wordpress.com/2014/07/10/makokoiwaya-waterfront-community-from-risk-to-resilience-through-regeneration-plan/.

67. Invest Online Real Estate Information and News Center, "Lagos to Building House Estate on the Lagoon," May 18, 2015, http://3invest.org.

68. "Lagos State Considers 'Makoko Floating School' into Development Plan," Archidatum: Architecture in Africa, May 15, 2015, http://archidatum.com/news/lagos-state-considers-makoko-floating-school-into-development-plan/.

69. Kerry H. Cook and Edward K. Vizy, "Impact of Climate Change on Mid-Twenty-first Century Growing Seasons in Africa," *Climate Dynamics* 39, no. 12 (2012): 2937–2955.

70. "Kenya Floods: Deadly Mosque Wall Collapse in Nairobi," *BBC News*, May 12, 2015, http://www.bbc.com/news/world-africa-32707150.

71. Chapter 6 features a more extensive discussion of the extraordinary community participation process associated with this project.

72. Chelina Odbert and Joe Mulligan, "The Kibera Public Space Project," in *Now Urbanism: The Future City Is Here, Change, Networked*, ed. Jeffrey Hou, Benjamin Spencer, Thaisa Way, and Ken Yocom (New York: Routledge, 2015), 177–192.

73. "India Heatwave: Death Toll Passes 2,500 as Victim Families Fight for Compensation," *The Telegraph*, June 2, 2015, http://www.telegraph.co.uk/news/worldnews/asia/india/11645731/India-heatwave-death-toll-passes-2500-as-victim-families-fight-for-compensation.html.

74. Stewart Gaffen and Cynthia Rosenzweig, *A Temperature and Seasonal Energy Analysis of Green, White, and Black Roofs* (New York: Center for Climate Systems Research, Columbia University, 2010).

75. Bureau of Energy Efficiency, *Energy Conservation Building Code User Guide* (New Delhi: Bureau of Energy Efficiency, 2009), 24–25.

76. Delhi Development Authority, *Delhi Street Design Guidelines* (New Delhi: Delhi Development Authority, 2009), 19–22.

77. In a separate rural initiative announced in 2014, India's Rural Development Ministry is planting two billion trees along the nation's highways—an undertaking that is also designed to address underemployment among youths; as many as 300,000 young people will be hired for the work. See Emily Thomas, "2 Billion New Trees to Be Planted in India, Official Says," *Huffington Post*, May 14, 2014, http://www.huffingtonpost.com/2014/06/14/india-tree-program-youth-jobs_n_5493583.html.

78. U.S. Environmental Protection Agency, "Reducing Urban Heat Islands: Compendium of Strategies," https://www.epa.gov/heatisland/resources/pdf/BasicsCompendium.pdf.

79. Ilham S. M. Elsayed, "Mitigation of the Urban Heat Island of the City of Kuala Lumpur, Malaysia," *Middle-East Journal of Scientific Research* 11, no. 11 (2012): 1602–1613, http://www.idosi.org/mejsr/mejsr11%2811%2912/19.pdf.

80. Royal Government of Cambodia, "Cambodia Post-Ketsana Disaster Needs Assessment," March 2010, http://www.adpc.net/igo/category/ID617/doc/2014-jNEq2U-ADPC-Cambodia_Post-Ketsana_PDNA.pdf.

81. Ibid.

82. Nordic Development Fund, "Rural Roads Improvement Project II," February 27, 2014, http://www.ndf.fi/project/rural-roads-improvement-project-ii-rrip-ii-ndf-c63.

83. Climate Investment Funds, "Cambodia: Provincial Roads Improvement Project—Climate Proofing of Roads in Prey Veng, Svay Rieng, Kampong Chhnang and Kampong Speu Provinces, under the Strategic Program for Climate Resilience for Cambodia," Meeting of the FIP Sub-committee, Washington, DC, November 2, 2011, Agenda Item 8, https://www-cif.climateinvestmentfunds.org/projects/provincial-roads-improvement-project-climate-proofing-roads-prey-veng-svay-rieng-kampong.

84. Nordic Development Fund, "Rural Roads Improvement Project II."

85. UN Water, "Climate Change Adaptation: The Pivotal Role of Water," UN Water Policy Brief, 1–18 at 1, http://www.unwater.org/downloads/unw_ccpol_web.pdf.

86. Jeremy Bird, "Water-Food-Energy Nexus," in *On Target for People and Planet: Setting and Achieving Water-Related Sustainable Development Goals*, ed. Julie Van Der Bliek, Peter McCornick, and James Clarke (Colombo, Sri Lanka: International Water Management Institute, 2014), 1–4 at 2, http://www.iwmi.cgiar.org/Publications/Books/PDF/setting_and_achieving_water-related_sustainable_development_goals.pdf.

87. Fred Pierce, "Pre-Inca Canals May Solve Lima's Water Crisis," *New Scientist*, April 9, 2015, https://www.newscientist.com/article/dn27311-pre-inca-canals-may-solve-limas-water-crisis/.

88. Gena Gammie and Bert De Bievre, "Assessing Green Interventions for the Water Supply of Lima, Peru: Cost Effectiveness, Potential Impact and Priority Research Areas," *Forest Trends* (2015): 5.

89. Armando Lamadri, "Climate Change, Adaptation, and Water in the Central Andes," in *Irrigation, Society, Landscape: Tribute to Thomas F. Glick*, ed. C. Sanchis-Ibor, G. Palau-Salvador, I. Mangue Alférez, and L. P. Martínez-Sanmartín (València: Universitat Politècnica de València, 2014), 820–832 at 822. doi:http://dx.doi.org/10.4995/ISL2014.2014.195.

90. Gammie and De Bievre, "Assessing Green Interventions," 4.

91. Ibid., 8.

92. Ibid., 7.

93. Ibid., 16.

94. Akmal Karimov, Vladimir Smakhtin, A. Mavlonov, and I. Gracheva, "Water 'Banking' in Fergana Valley Aquifers—A Solution to Water Allocation in the Syrdarya River Basin?" *Agricultural Water Management* 97, no. 10 (2010): 1461–1468. The Aral Sea, formerly one of the world's four largest seas, has nearly disappeared since rivers were diverted by Soviet irrigation projects, and it has since become heavily polluted, contributing to public health problems.

95. Bird, "Water-Food-Energy Nexus," 1–4.

96. International Water Management Institute (IWMI), "Banking on Groundwater in Times of Change," IWMI Water Policy Brief no. 32 (Colombo, Sri Lanka: International Water Management Institute, 2010): 1–8 at 5. doi:10.3910/2009.203.

97. Ibid.

98. Nuria Font and Joan Subirats, "Water Management in Spain: The Role of Policy Entrepreneurs in Shaping Change," *Ecology and Society* 15, no. 2 (2010), http://www.ecologyandsociety.org/vol15/iss2/art25/.

99. Blanca Jiménez, Pay Drechsel, Doulaye Koné, Akiça Bahri, Liqa Raschid-Sally, and Manzoor Qadir, "Wastewater, Sludge and Excreta Use in Developing Countries: An Overview," in *Wastewater Irrigation and Health: Assessing and Mitigation Risk in Low-Income Countries*, ed. Pay Drechsel, Christopher A. Scott, Liqa Raschid-Sally, Mark Redwood, and Akiça Bahri (London: Earthscan, 2010), 1–400, http://publications.iwmi.org/pdf/h042601.pdf.

100. Intizar Hussain, Liqa Raschid, Munir A. Hanjra, Fuard Mrikar, and Wim van der Hoek, "Wastewater Use in Agriculture: Review of Impacts and Methodological Issues in Valuing Impacts," International Water Management Institute Working Paper 37, Colombo, Sri Lanka, International Water Management Institute, 2002.

101. Ibid., 3.

102. Ibid., 7.

103. Anton Mauderli and Otto Schreiner, "Use of Reclaimed Water in the Jordan Valley: Appraisal Mission Final Report," Deutsche Gesellschaft für Technische Zusammenarbeit GmbH (GTZ), September 2002, 1–23 at 8.

104. A. Vallentin, J. Schlick, F. Klingel, P. Bracken, and C. Werner, "Case Study of Sustainable Sanitation Projects: Use of Treated Wastewater in Agriculture, Jordan Valley, Jordan," Sustainable Sanitation Alliance (SuSanA), http://www.susana.org/en/resources/case-studies/details/78.

105. Bashaar Y. Ammary, "Wastewater Reuse in Jordan: Present Status and Future Plans," *Desalination* 211 (2007): 164–176 at 166.

106. Vallentin et al., "Sustainable Sanitation Projects."

107. Ibid.

108. Ibid.

109. Sameer Abdel-Jabbar, "Reuse of Treated Wastewater in the Jordan Valley: An Added Value Outweighs Risks," Gesellschaft für Technische Zusammenarbeit GmbH (GTZ) document, http://www.susana.org/_resources/documents/default/2-78-en-susana-cs-jordan-treated-wastewater-reuse-2009.pdf. In addition to being more costly, chemical fertilizers are associated with the risk of soil salinization.

110. A. M. Manion, *Carbon and Its Domestication* (Dordrecht: Springer, 2006), 190.

111. Ibid.

112. Valletin et al., "Sustainable Sanitation Projects."

113. Richard T. Conant, "Challenges and Opportunities for Carbon Sequestration in Grassland Systems: A Technical Report on Grassland Management and Climate Change Mitigation," *Integrated Crop Management* 9 (2010): 1–57 at 3.

114. Constance Elizabeth Hunt, *Thirsty Planet: Strategies for Sustainable Water Management* (Chicago: Zed Books and University of Chicago Press, 2004), 91.

115. S. R. Carpenter and R. Biggs, "Freshwaters: Managing across Scales in Space and Time," in *Principles of Ecosystem Stewardship*, ed. F. S. Chapin, G. P. Konas, and C. Folke (New York: Springer, 2010), 197–220.

116. Allan Savory, "A Global Strategy for Addressing Global Climate Change," The Soil Carbon Coalition, 1–20 at 11, http://soilcarboncoalition.org/files/globalstrategy.pdf.

117. Ibid., 2.

118. Allan Savory, "Holistic Resource Management: A Conceptual Framework for Ecologically Sound Economic Modeling," *Ecological Economics* 3 (1991): 181–191 at 186.

119. Andrea Malberg, "Africa Centre for Holistic Management: Case Study," Savory Institute, 1–9 at 4, http://savoryinstitute.net/media/41141/ACHM-long-version_24apr.pdf.

120. Ibid., 7–8.

121. Savory, "Global Strategy for Addressing Global Climate Change."

122. Savory Institute, "Restoring the Climate through Capture and Storage of Soil Carbon through Holistic Planned Grazing," adapted from A. D. Sacks, R. Teague, F. Provenza, J. Laurie, S. Itzkan, and K. Thidemann, "Re-establishing the Evolutionary Grassland-Grazer Relationship for Long-Term Sequestration of Carbon in Soils: Restoring Atmospheric Carbon Dioxide to Pre--industrial Levels," in *Geotherapy: Innovative Methods of Soil Fertility Restoration, Carbon Sequestration, and Reversing CO$_2$ Increase*, ed. Thomas J. Goreau, Ronal W. Larson, and Joanna Campe (Boca Raton, FL: CRC Press and Taylor and Francis Group, 2014), 1–20 at 11, http://savory.global/assets/docs/evidence-papers/restoring-the-climate.pdf.

123. Ibid., 14.

124. Afreen Siddiqi and Laura Diaz Anadon, "The Water-Energy Nexus in Middle East and North Africa," *Energy Policy* 39 (2011): 4529–4540 at 4530.

125. The World Bank, *Renewable Energy Desalination: An Emerging Solution to Close the Water Gap in the Middle East and North Africa* (Washington, DC: The World Bank: The International Bank for Reconstruction and Development, 2012), 29.

126. Thermal desalination uses energy to evaporate seawater and then condense it as fresh. Reverse osmosis pretreats seawater to remove solids and then forces the water through a membrane, inhibiting the passage of salt.

127. Paul Vidal de la Blache and Stacy Kourlis, "Water and Energy Nexus: A Literature Review," *Water in the West Publications* (Stanford, CA: Stanford University, 2013), 1–136 at 13, http://waterinthewest.stanford.edu/sites/default/files/Water-Energy_Lit_Review_0.pdf.

128. International Desalination, http://idadesal.org/desalination-101/desalination-by-the-numbers/.

129. K. Quteishat, "Desalination and Water Affordability," presented at SITeau International Conference, Casablanca, Morocco, January 2009.

130. Siddiqi and Anadon, "Water-Energy Nexus," 4538.

131. Lourdes Garcia-Rodriguez, Ana I. Palmero-Marrero, and Carlos Gomez-Camacho, "Comparison of Solar Thermal Technologies for Applications in Seawater Desalination," *Desalination* 142, no. 2 (2002): 135–142.

132. Giles Parkinson, "Saudis to Build World's First Large Scale Solar Powered Desalination Plant," ReNew Economy, January 22, 2015, http://reneweconomy.com.au/2015/saudis-build-worlds-first -large-scale-solar-powered-desalination-plant-82903.

133. Patrick McCully, "Chile Has the Most Extensive Energy Resources in the World," This Is Chile, Environment News, June 1, 2012, https://www.thisischile.cl/chile-has-the-most-extensive -energy-resources-in-the-world/?lang=en.

134. Ibrahim Al-Mutaz and M. I. Ahmed, "Evaluation of Solar Powered Desalination Processes," *Desalination* 73 (1989): 181–190 at 181. doi:10.1016/0011-9164(89)87012-2.

135. Eleonora Cominelli, Massimo Galbiati, Chiara Tonelli and Chris Bowler, "Water: The Invisible Problem," *EMBO Reports* 10, no. 7 (2009): 661–676 at 673.

136. Ibid., 674.

137. Charlie Paton, "Vapour Value," in *Agriculture, Water Security*, Global Water Forum, April 1, 2014, http://www.globalwaterforum.org/2014/04/01/vapour-value/.

138. The Sahara Forest Project, "The Sahara Forest Project, Qatar," 1–24 at 5, http://saharaforest project.com/qatar/.

139. The Sahara Forest Project, "Enabling Restorative Growth," 1–20 at 8, https://cpanel46.proisp .no/~saharoxf/wp-content/uploads/2015/03/SFP-brosjyre-20-sider_2014_low-22102014.pdf.

140. Ibid.

141. Charlie Patton and Philip Davies, "The Seawater Greenhouse: Cooling, Freshwater and Fresh Produce from Seawater," *Proceedings from the 2nd International Conference on Water Resources in Arid Environments*, Riyadh, 2006, 1–21 at 13.

142. Daniel Clery, "Greenhouse-Power Plant Hybrid Set to Make Jordan's Desert Bloom," *Science* 331 (January 14, 2011): 136, http://www.sciencemag.org/content/331/6014/136.

143. Sahara Forest Project, "EU Grant to Build in Jordan," Press Release, May 28, 2015, http:// saharaforestproject.com/eu-grant-build-jordan/.

144. Clery, "Greenhouse-Power Plant," 136.

145. Tina Casey, "Sahara Forest Project Grows Food and Biofuel," Clean Technica, April 19, 2014, http://cleantechnica.com/2014/04/19/sahara-forest-project-grows-food-biofuel/.

146. The Sahara Forest Project, "The Sahara Forest Project, Qatar," 1–24 at 16, http://saharaforestproject .com/qatar/.

147. Patton and Davies, "Seawater Greenhouse," 2.

Chapter 6

1. D. W. Gade, "Bridge Types in the Central Andes," *Annals of the Association of American Geographers* 62, no. 1 (March 1972): 94–109 at 96.

2. Ibid.

3. Ibid., 96.

4. Loren McIntyre, "The Lost Empire of the Incas," *National Geographic* 144, no. 6 (1973): 729–787.

5. Ric Finch, "Keshwa Chaca: Straw Bridge of the Incas," *South American Explorer*, no. 69 (Fall–Winter 2002): 7–13 at 9.

6. Ibid.

7. Richard Heinberg, *The End of Growth: Adapting to Our New Economic Reality* (Gabriola Island, BC, Canada: New Society Publishers, 2011), 223.

8. Term coined in a 1969 article by Sherry R. Arnstein, "A Ladder of Citizen Participation," *American Institute of Planning Journal* 35, no. 4 (July 1969): 216–224.

9. Ghazal Mansuri and Vajayendra Rao, *Localizing Development: Does Participation Work?* (Washington, DC: World Bank, 2013), 28.

10. Ibid., 2.

11. The World Bank, "Implementation Completion and Results Report (IDA-35320) on a Credit in the Amount of SDR 81.9 Million to the Socialist Republic of Vietnam for a Community Based Rural Infrastructure Project," December 28, 2009, 1–63 at 1.

12. Ibid., 6.

13. Tosca Van Vijfejken, "Empowering Poor Communities through Decentralized Decision Making: The Vietnam Community Based Rural Infrastructure Project," Social Development Notes no. 66, Social Development Family in the Environmentally and Socially Sustainable Development Network of the World Bank (Washington, DC: World Bank, 2001), 1–4 at 2.

14. The World Bank, "Implementation Completion and Results Report," 6.

15. World Bank, "Vietnam: Community Based Rural Infrastructure Project," News and Broadcast, February 7, 2012, http://web.worldbank.org/WBSITE/EXTERNAL/NEWS/0,,contentMDK:2288318 4~menuPK:141310~pagePK:34370~piPK:34424~theSitePK:4607,00.html.

16. World Bank, "Implementation Completion and Results Report (IDA-35320) on a Credit in the Amount of SDR 81.9 Million (US$ 102.78 Million Equivalent) to the Socialist Republic of Vietnam for a Community Based Rural Infrastructure Project, December 28, 2009," 11.

17. World Bank, "Rural Infrastructure Project."

18. Kirk Herbertson, Athena Ballesteros, Robert Goodland, and Isabel Munilla, *Breaking Ground: Engaging Communities in Extractive and Infrastructure Projects* (Washington, DC: World Resources Institute, 2009), 1–47 at 8, http://pdf.wri.org/breaking_ground_engaging_communities.pdf.

19. Empresas Públicas de Medellín (EPM), "Jepírachi: An Experience with the Wayuu Indigenous Community from the Upper Guajira in Colombia" (Medellín: EPM, 2010), 1–72 at 8, https:// www.epm.com.co/site/documentos/mediosdecomunicacion/publicacionesimpresas/jepirachi /LibroJepirachieningles.pdf.

20. Oliver Balch, "Colombia's Cerrejón Mine: The Social Impact on Surrounding Communities," *The Guardian*, July 25, 2013, https://www.theguardian.com/sustainable-business/colombia -cerrejon-mine-social-impact-communities.

21. United Nations Framework Convention on Climate Change Clean Development Mechanism, Project Design Document Project 0194: Jepirachi Wind Power Project, 1–61 at 5, https:// wbcarbonfinance.org/docs/Jepirachi_PDD.pdf.

22. Walter Vergara, Alejandro Deeb, Natsuko Toba, Peter Cramton, and Irene Leino, "Wind Energy in Colombia: A Framework for Market Entry" (Washington, DC: The World Bank, 2010), i–120 at xv, http://www.cramton.umd.edu/papers2010-2014/vergara-deep-toba-cramton-leino-wind-energy -in-colombia.pdf.

23. Empresas Públicas de Medellín, "Jepírachi," 12.

24. George Ledec, Kenna Rapp, and Roberto Aiello, "Greening the Wind: Environmental and Social Considerations for Wind Power Development" (Washington, DC: The World Bank: The International Bank for Reconstruction and Development, 2011), 1–172 at 107, http://documents .worldbank.org/curated/en/239851468089382658/Greening-the-wind-environmental-and-social -considerations-for-wind-power-development.

25. The World Bank, Projects and Operations, "Jepirachi Carbon Offset Project," http://www .worldbank.org/projects/P074426/jepirachi-carbon-off-set-project?lang=en.

26. Empresas Públicas de Medellín, "Jepírachi," 20–25.

27. Ibid., 49.

28. Ibid., 57.

29. Ledec et al., "Greening the Wind," 110.

30. Empresas Públicas de Medellín, "Jepírachi," 63.

31. Ibid., 8.

32. Larry Greenemeier, "Winds of Change Blow Renewable Energy across Latin America," *Scientific American On-Line*, July 2, 2010, http://www.scientificamerican.com/article/wind-power -colombia-guajira/.

33. Ibid.

34. Fredrick Ochieng, "Enhanced Energy Access for the Urban Poor: Best Practice Casebook," UN Habitat, http://mirror.unhabitat.org/downloads/docs/Enhanced%20Energy%20Access%20for %20Urban%20poor%20Best%20Practice%20Case%20book.pdf.

35. Jean-Elie Aron, Olivier Kayser, Larent Liautaud, and Aileen Nowland, "Access to Energy for the Base of the Pyramid," Ashoka, October 2009, 1–100 at 11, https://www.ashoka.org/sites/ashoka /files/Ashoka-HYSTRA_Access_to_%20Energy_for_the_BOP.pdf.

36. Kennedy Muzee, "Global Energy Network for Urban Settlements (GENUS)," *GENUS News-letter*, no. 4 (2012): 1–8 at 2, http://www.urbangateway.org/urbanyouth/document/genus-newsletter -issue-4.

37. Connie Smycer, "Promoting Energy Access for the Urban Poor in Africa: Approaches and Challenges in Slum Electrification," background paper prepared for the workshop of the same title, Nairobi, Kenya, October 26–27, 2009, 1–30 at 25.

38. Aron et al., "Access to Energy," 24.

39. Smycer, "Promoting Energy Access," 25.

40. Ochieng, "Enhanced Energy Access," 17.

41. Emani Kumar, "Operationization of Urban Nexus Approaches in Nashik, India," ICLEI Global, PowerPoint presentation, May 29, 2014, Slide #7, http://www.waternexussolutions.org /contentsuite/upload/wns/all/Session%201_5%20Emani%20Kumar.pdf.

42. Ritu Thakur and Emani Kumar, "Case Study, Nashik, India: Demonstrating the Urban NEXUS Approach to Optimize Water, Energy and Land Resources in Peri-urban Agriculture," GIZ, August 2014, http://www2.giz.de/wbf/4tDx9kw63gma/02_UrbanNEXUS_CaseStudy_Nashik.pdf.

43. Kumar, "Operationization of Urban Nexus," Slide #7.

44. Thakur and Kumar, "Demonstrating the Urban NEXUS Approach."

45. Kumar, "Operationization of Urban Nexus," Slide #4.

46. Thakur and Kumar, "Demonstrating the Urban NEXUS Approach."

47. Kumar, "Operationization of Urban Nexus," Slide #22.

48. ICLEI website, Case Studies, "Demonstrating the Urban NEXUS Approach to Optimize Water, Energy and Land Resources in Peri-Urban Agriculture," http://www.iclei.org/fileadmin /PUBLICATIONS/Case_Studies/Urban_NEXUS_cs02_Nashik_ICLEI-GIZ_2014.pdf.

49. Ibid., Slide #5.

50. Ibid., Slide #10.

51. Ibid., Slide #4.

52. Ibid., Slide #20.

53. Ibid., Slide #10.

54. Saleem Shaikh and Sughra Tunio, "Mangroves Reduce Disaster Risk, Boost Incomes in Vietnam," Thomas Reuters Foundation, May 14, 2012, http://www.trust.org/item/?map=mangroves
-reduce-disaster-risk-boost-income-options-in-vietnam/.

55. Pauline Buffle, Nguyen Thi Yen, and Morten Fauerby Thomsen, "Community-Based Mangrove Reforestation and Management in Da Loc, Vietnam," Ecosystem and Livelihoods Adaptation Networks (ELAN), 2001, 1–11, http://www.preventionweb.net/files/25381_vietnam.pdf.

56. Nguyen Hoang Tri, W. N. Adger, and M. M. Kelly, "Natural Resource Management in Mitigating Climate Impacts: The Example of Mangrove Restoration in Vietnam," *Global Environmental Change* 8, no. 1 (1998): 49–61 at 50.

57. Each household had one vote.

58. Buffle et al., "Community-Based Mangrove Reforestation and Management," 54.

59. Ibid., 57.

60. Ibid., 55.

61. CARE, "International Climate Change Brief: Adaptation, Gender and Women's Empowerment," http://www.care.org/sites/default/files/documents/CC-2010-CARE_Gender_Brief.pdf.

62. Shaikh and Tunio, "Mangroves."

63. Buffle et al., "Community-Based Mangrove Reforestation and Management," 59.

64. Ibid.

65. Malick Gaye and Fode Diallo, "Community Participation in the Management of the Urban Environment in Rufisque (Senegal)," *Environment and Urbanization* 9, no. 1 (April 1997): 9–29 at 15.

66. United Nations Habitat, "Community Participation in the Enhancement of a Sustainable Environment, Dakar, Senegal. Best Practices Database," http://www.ucl.ac.uk/dpu-projects/drivers_urb
_change/urb_infrastructure/pdf_city_planning/HABITAT_BestPractice_Community_Dakar.pdf.

67. Ibid., 20.

68. ENDA Tiers Monde, "PADE Study," August 1997, www.globenet.org/preceup/angl/docs_angl
/PADE%20Study.rtf.

69. United Nations Habitat, "Community Participation."

70. ENDA Tiers Monde, "PADE Study."

71. United Nations Habitat, "Community Participation."

72. Ibid.

73. Martin Medina, "The Informal Recycling Sector in Developing Countries: Organizing Waste Pickers to Enhance Their Impact," *Grid Lines*, Publication of PPIAF (Public-Private Infrastructure,

Advisory Facility), World Bank, Note no. 44, October 2008, 1–4 at 1, https://www.ppiaf.org/sites /ppiaf.org/files/publication/Gridlines-44-Informal%20Recycling%20-%20MMedina.pdf.

74. Ana Paula Bortoleto and Keisuke Hanaki, "Analysis of Waste Prevention Policies Applied at the Solid Waste Management in Belo Horizonte, Brazil," International Solid Waste Association, 1–8 at 4, http://www.iswa.org/uploads/tx_iswaknowledgebase/543801_Paper.pdf.

75. Sonia M. Dias, "Belo Horizonte Selective Waste Collection Programme—Social Inclusion Features," as drawn from a presentation at WATSAN 2010, UN Habitat, 2010, 1–8 at 2, WIEGO, http:// wiego.org/sites/wiego.org/files/publications/files/Dias_BELO_HORIZONTE_SELECTIVE_WASTE _COLLECTION_PROGRAMME.pdf.

76. Ibid., 3.

77. Ibid.

78. LSE Cities, ICLEI Local Governments for Sustainability, and Global Green Growth Institute, "Going Green—How Cities Are Leading the Next Economy," 2012, 1–115 at 78, https://files .lsecities.net/files/2013/06/Going-Green-Final-Edition-web-version.pdf.

79. Charles Landry, *The Creative City: A Toolkit for Urban Innovators*, second edition (London: Earthscan, 2008), 151.

80. LSE Cities, ICLEI Local Governments for Sustainability, and Global Green Growth Institute, "Going Green," 80.

81. Sonia M. Dias, "Integrating Waste Pickers for Sustainable Recycling," paper presented at international workshop Planning for Sustainable and Integrated Waste Management, Manila, September 18–21, 2000, 1–9 at 3, Waste Portal, http://wasteportal.net/en/system-elements/collection /integrating-waste-pickers-sustainable-recycling.

82. LSE Cities, ICLEI Local Governments for Sustainability, and Global Green Growth Institute, "Going Green," 80.

83. Medina, "Informal Recycling Sector," 4.

84. Population according to a 2005 census.

85. International Partnership for the Satoyama Initiative, "Forest Management through Community-Based Forest Enterprises in Ixtlan de Juarez, Oaxaca, Mexico," Case Study, March 5, 2010, http:// satoyama-initiative.org/en/forest_management_through_community-based_forest_enterprises_in _ixtlan_de_juarez/.

86. Although most of Ixtlán's forests are located in the highlands of the Sierra Norte, farther down the mountain slopes, the community-protected area also includes tropical forests. See Peter Gill, "Community Forestry in Oaxaca: An Assessment of the Level of Autonomy in Forest Decision-Making and Its Environmental and Economic Importance," Digital Collections, School for International Training, ISP Collection, December 2007, 3–29 at 24, http://digitalcollections.sit.edu/cgi /viewcontent.cgi?article=1181&context=isp_collection.

87. James S. Gruber, "Perspectives of Effective and Sustainable Community-Based Natural Resource Management: An Application of Q Methodology to Forest Projects," *Conservation and Society* 9, no. 2 (2011): 159–171 at 162.

88. David Barton Bray, "Capitalism Meets Common Property," from "Voices from the New Generation" issue, *Americas Quarterly* (Winter 2010): 31–35 at 34.

89. Gruber, "Perspectives of Effective and Sustainable Community-Based Natural Resource Management."

90. Bray, "Capitalism," 33.

91. Elisabeth Malkin, "Growing a Forest, and Harvesting Jobs," *New York Times*, November 22, 2010.

92. Gill, "Community Forestry," 22.

93. Ibid.

94. Bray, "Capitalism," 35.

95. L. Gomez-Mendoza, E. Vega-Peña, M. I. Ramírez, J. P. Placio-Prieto, and L. Galicia, "Projecting Land-Use Change Processes in the Sierra Norte of Oaxaca, Mexico," *Applied Geography* 26 (2006): 276–290. A secondary benefit of reforestation is that it offsets the effects of erosion, which reduces the amount of carbon that can be stored in soil.

96. Silviculture is the art and science of controlling the establishment, growth, composition, and quality of forest vegetation for the full range of forest resource objectives.

97. Talli Nauman, "Mexican Forest Communities See Business Benefit from Climate Laws," Thomas Reuters Foundation, June 5, 2014, http://www.trust.org/item/20140605132542-qj7ow/.

98. Garret Hardin's theory, a theory propounded by governments to justify the takeover of forest management.

99. Bray, "Capitalism," 32.

100. Ibid., 35.

101. "Keeping It in the Community: Well-Organised Locals Often Make the Best Forest Managers, But They Need Help," *The Economist*, Special Reports: Forests, September 23, 2010 print edition, http://www.economist.com/node/17062703/print.

102. United Nations, "Innovation for Sustainable Development: Local Case Studies from Africa," 1–62 at 21, https://sustainabledevelopment.un.org/content/documents/publication.pdf.

103. The remainder of the cost was funded by the Roman Catholic Church (10 percent) and the United Nations Development Programme (42 percent). See ibid., 22.

104. United Nations Development Programme, "Community Water Initiative: Project Portfolio Review," November 1, 2006, 1–23 at 16, http://www.undp.org/content/undp/en/home/librarypage

/environment-energy/water_governance/community-water-initiative-project-portfolio-review
.html.

105. Ibid.

106. United Nations Development Group, "Millennium Development Goals Good Practices, Chapter 4: MDG-7: Environmental Sustainability," 1–90 at 83.

107. United Nations, "Innovation for Sustainable Development," 21.

108. United Nations Development Group, "Millennium Development Goals," 15.

109. United Nations, "Innovation for Sustainable Development," 23.

110. Z. Aycan, "Leadership and Teamwork in Developing Countries: Challenges and Opportunities," in *Online Readings in Psychology and Culture*, ed. W. J. Lonner, D. L. Dinnel, S. A. Hayes, and D. N. Sattler (Bellingham: Center for Cross-Cultural Research, Western Washington University, 2011), Unit 15, Chapter 8, 1–13 at 4.

Chapter 7

1. Muhammad Yunus and Judith Rodin, "Save the World, Turn a Profit," *Bloomberg View*, September 25, 2015.

2. United Nations Framework Convention on Climate Change, "Clean Development Mechanism Definition," http://unfccc.int/kyoto_protocol/mechanisms/items/1673.php.

3. Fiona Harvey, "Global Carbon Trading System Has 'Essentially Collapsed,'" *The Guardian*, September 10, 2012, https://www.theguardian.com/environment/2012/sep/10/global-carbon-trading
-system.

4. Carbon Finance Unit, World Bank Climate Policy and Finance Department, "Options for Streamlining the CDM Project Cycle" (Washington, DC: World Bank, 2014).

5. Peter Reina, "Dual Purpose Tunnel Becomes a Very Smart Solution: Putting Road into Malaysia Water Tunnel Is Key to Affordability," April 25, 2005, http://enr.construction.com/features
/transportation/archives/050425-1.asp.

6. Shahriyar Anam, Abu Saleh Md. Sohel-Uz-Zaman, and Iftekhar Anam, "Impact of Jamuna Multipurpose Bridge on the Local Economy," IABSE Symposium Report 89, no. 1 (Zurich: International Association for Bridge and Structural Engineering, 2005).

7. Justin Guay, "Meet Uttar Pradesh, India's Next Distributed Solar Hotbed," *Sierra Club, India Environment Post*, November 19, 2013, http://action.sierraclub.org/site/MessageViewer?em_id=265265
.0&dlv_id=224822.

8. Julien Bouissou, "Indian City of Surat Anticipates Worst Effects of Climate Change," *The Guardian*, September 15, 2014, https://www.theguardian.com/cities/2014/sep/15/indian-cities-climate
-change-surat.

9. "100 Resilient Cities," Rockefeller Foundation, https://www.rockefellerfoundation.org/our -work/initiatives/100-resilient-cities/.

10. Asian Development Bank, "Establishment of the Urban Climate Change Resilience Trust Fund under the Urban Financing Partnership Facility," December 2013, www.adb.org/documents /establishment-urban-climate-change-resilience-trust-fund-under-ufpf.

11. Norfund Investments, "Kinangop Wind Park," http://www.norfund.no/eastern-africa/kinangop -wind-park-article1027-319.html.

12. Norman Aquino, "Manila Water to Work on Myanmar Project with Mitsubishi," Bloomberg, March 17, 2014, http://www.bloomberg.com/news/articles/2014-03-17/manila-water-to-develop -project-in-myanmar-to-cut-water-leaks.

13. Egypt PPP Law (law no. 67, 2010), http://www.pppcentralunit.mof.gov.eg.

14. "Jamuna Bridge Connects Bangladesh's Two Halves," World Bank, http://web.worldbank.org /WBSITE/EXTERNAL/EXTABOUTUS/IDA/0,,contentMDK:21271555~menuPK:3266877~pagePK:5 1236175~piPK:437394~theSitePK:73154,00.html.

15. "Getting Approval from Multiple Agencies Biggest Hurdle," *The Times of India*, October 1, 2015, http://timesofindia.indiatimes.com/city/delhi/Getting-approval-from-multiple-agencies-biggest -hurdle/articleshow/49174655.cms.

16. Reina, "Dual Purpose Tunnel."

17. Shiv Kumar, "A First—Solar Plant on Canal to Generate Power," *The Tribune India*, January 25, 2015, http://www.tribuneindia.com/news/sunday-special/kaleidoscope/a-first-solar-plant-on-canal -to-generate-power/34021.html.

18. Maina Waruru, "Gusts of Opposition Hit Kenyan Wind Farm Project," Reuters, May 21, 2015, http://www.reuters.com/article/2015/05/21/kenya-windpower-protests-idUSL5N0YB4ID2015 0521.

19. David Hunter, "Using the World Bank Inspection Panel to Defend the Interests of Project-Affected People," *Chicago Journal of International Law* 4, no. 1 (2003): 201–211.

20. Virendra Pandit, "Now, Gujarat to Cover Narmada Canals with Solar Panels!" *The Hindu Business Line*, April 23, 2012, http://www.thehindubusinessline.com/economy/policy/now-gujarat-to -cover-narmada-canals-with-solar-panels/article3346191.ece.

21. Arab Countries Water Utility Association, *Wastewater Reuse in Arab Countries* (Amman: Arab Countries Water Utility Association, 2010).

22. Cecilia Tortajada, "Water Management in Singapore," *Water Resources Development* 22, no. 2 (2006): 227–240.

23. Marco Arnone and Leonardo Borlini, *Corruption: Economic Analysis and International Law* (Northampton, MA: Edward Elgar Publishing, 2014), 47–52.

24. Ibid.

25. United States Department of Justice, Foreign Corrupt Practices Act, https://www.justice.gov/criminal-fraud/foreign-corrupt-practices-act.

26. Joseph W. Yockey, "Solicitation, Extortion, and the FCPA," *Notre Dame Law Review* 87 (2011): 781.

Chapter 8

1. Broadband cables are affixed to the electrical transmission and distribution towers.

2. Deloitte LLP, "Unlocking Broadband for All: Broadband Infrastructure Sharing Policies and Strategies in Emerging Markets," Association for Progressive Communications, April 2015, 1–86 at 5, https://www.apc.org/en/system/files/Unlocking%20broadband%20for%20all%20Full%20report.pdf.

3. Ibid., 2.

4. Ibid., 6.

5. David W. Orr, "Systems Thinking and the Future of Cities," *The Solutions Journal* 5, no. 1 (May 16, 2014): 53–61.

6. The recommendations under this heading draw on the following sources: Deloitte LLP, "Unlocking Broadband for All," 51–56; and Jeb Brugmann with Katherine Brekke and Lucy Price, "Operationalizing the Urban NEXUS: Towards Resource-Efficient and Integrated Cities and Metropolitan Regions" (Bonn: Deutsche Gesellschaft für Internationale Zusammenarbeit GmbH and ICLEI, 2014), 1–104, http://resilient-cities.iclei.org/fileadmin/sites/resilient-cities/files/Full_papers/Urban_NEXUS_Publication_ICLEI-GIZ_2014_web.pdf.

7. Peter H. Gleick, "Global Freshwater Resources: Soft-Path Solutions for the 21st Century," *Science* 302, no. 5650 (November 14–28, 2003): 1524–1528 at 1527.

8. Refer to the comprehensive definitions and practices defined by the Global Water Partnership, http://www.gwp.org/The-Challenge/What-is-IWRM/.

9. United Nations World Water Assessment Programme (WWAP), *The United Nations World Water Development Report 2014: Water and Energy* (Paris: UNESCO, 2014), 7.

10. Mattia Romani, James Rydge, and Nicholas Stern, "Recklessly Slow or Rapid Transition to a Low-Carbon Economy? Time to Decide," Centre for Climate Change Economics and Policy and Grantham Research Institute on Climate Change and the Environment, December 2012, 1–15 at 3.

11. Ibid., 7. Module prices have dropped from over $100/W in the 1970s, to around $2/W in 2010, and to around $0.90/W in 2012.

12. Ibid., 4.

13. PV Magazine, "Ghana Finalizes Plans for 155 MW Solar Park," March 4, 2014, http://www.pv-magazine.com/news/details/beitrag/ghana-finalizes-plans-for-155-mw-solar-park_100014400/#axzz3jGOa7LSD.

14. Janet L. Sawin and Freyr Sverrisson, "Renewables 2014: Global Status Report," Renewable Energy Network for the 21st Century, 1–216 at 93–94, http://www.ren21.net/Portals/0/documents /Resources/GSR/2014/GSR2014_full%20report_low%20res.pdf.

15. Ibid., 96.

16. Lester R. Brown, "On Rooftops Worldwide a Solar Heating Revolution," Earth Policy Institute, http://www.earth-policy.org/book_bytes/2010/pb4ch05_ss3a.

17. Carlos Faria, "The Use of Thermal Solar Energy in Brazil," *Alternative Energy Magazine*, December 1, 2006, http://www.altenergymag.com/content.php?post_type=1487.

18. Ansel Eisentraut and Adam Brown, "Heating without Global Warming: Market Developments and Policy Considerations for Renewable Heat" (Paris: International Energy Agency Publications, April 2014), 1–92 at 59, https://www.iea.org/publications/freepublications/publication /heating-without-global-warming.html.

19. Ibid., 61.

20. Brian Guzzone and Amy Alexander, "The Importance of Landfill Gas Energy in Integrated Municipal Solid Management in the Developing World," *Waste Advantage Magazine*, August 15, 2014, http://wasteadvantagemag.com/the-importance-of-landfill-gas-energy-in-integrated-munici pal-solid-management-in-the-developing-world/.

21. Chris Godlove, "Overcoming Barriers to Landfill Gas Energy Recovery," presentation, 16th Annual Landfill Methane Outreach Program Conference, June 27, 2013, https://www3.epa.gov /lmop/documents/pdfs/conf/16th/09b_Godlove_presentation.pdf.

22. United Nations Asian and Pacific Centre for Agricultural Engineering and Machinery (UNAP-CAEM), "Recent Development in Biogas Technology for Poverty Reduction and Sustainable Development" (Beijing: United Nations Economic and Social Commission for Asia and the Pacific [UNESCAP], 2007), 55, http://www.un-csam.org/publication/F-Biogas.PDF.

23. Ibid., 59.

24. The World Bank, *World Development Report 2010: Development and Climate Change* (Washington, DC: The World Bank: The International Bank for Reconstruction and Development, 2010), 330–331.

25. Lily Ryan-Collins, Karen Ellis, and Alberto Lemma, "Climate Compatible Development in the Infrastructure Sector," Engineers against Poverty and the Overseas Development Institute (ODI), UK Institution of Civil Engineers, June 2011, 15.

26. Pranab Bardhan and Dilip Mookherjee, "Delivering Public Services in Developing Countries: How the Poor Can Benefit from Decentralisation," *Economic Journal*, no. 508 (January 2006): 101–127, http://www.res.org.uk/details/mediabrief/4387151/Delivering-Public-Services-In-Developing -Countries-How-The-Poor-Can-Benefit-From.html.

27. Hillary Brown, *Sun, Sea, Salt and Soil: A "Circular Economy" for Anse-Rouge, Haiti* (unpublished manuscript, The City College of New York, 2015).

28. The project was developed by the author with her interdisciplinary graduate students at the City College of New York as a short study project for the Case Studies in Sustainability class, an elective in the "Sustainability in the Urban Environment," Master of Science program.

29. Ibid.

30. David W. Orr, *The Nature of Design: Ecology, Culture and Human Intention* (Oxford: Oxford University Press, 2002), 20.

Acronyms

ACHM: Africa Centre for Holistic Management
ADB: Asian Development Bank
AFI: artificial floating island
AIIB: Asian Infrastructure Investment Bank
ASCE: American Society of Civil Engineers
ASMARE: Associação dos Catadores de Papel, Papelão e Material Reaprovitável (Waste Pickers' Association, Belo Horizonte, Brazil)
C40: C40 Cities Climate Leadership Group
CARE: Cooperative for Assistance and Relief Everywhere
CDM: Clean Development Mechanism (United Nations project)
CER: certified emission reduction
CMMB: Community-Based Mangrove Management Board
CPCC: Commune Project Coordinating Committee
CSIR: Council of Scientific and Industrial Research
CSP: concentrated solar power
CTRS: Centro de Tratamiento de Residuos Solidos (Treatment Center for Solid Waste, Brazil)
CWSSP: Community Water Supply and Sanitation Project (USAID)
ECBC: Energy Conservation Building Code
EIP: ecoindustrial park
EKW: East Kolkata Wetlands
EPA: Environmental Protection Agency (U.S.)
EPM: Empresas Públicas de Medellín

EU: European Union

FCPA: Foreign Corrupt Practices Act (U.S.)

FOCAUP: Fonds Communautaire pour l'Assainissement des quartiers Urbains Pauvres (Community Fund for Sanitation in Poor Urban Neighborhoods)

GDP: gross domestic product

GEF: Global Environment Facility

GHG: greenhouse gas

GIZ: Gesellschaft für Internationale Zusammenarbeit GmbH

GTZ: Deutsche Gesellschaft für Technische Zusammenarbeit GmbH

IB: Itaipu Binacional

IBC: International Building Code

ICLEI: International Council for Local Environmental Initiatives

ICT: information and communications technology

IDB: Inter-American Development Bank

IFAD: International Fund for Agricultural Development

IIED: International Institute for Environmental Development

IMF: International Monetary Fund

IPCC: Intergovernmental Panel on Climate Change

ISO: International Standards Organization

ISWM: integrated solid waste management

ITP: Itaipu Technological Park

IUPR: Institute for Urban Planning Research (Curitiba, Brazil)

IWMI: International Water Management Institute

IWRM: Integrated Water Resources Management

JVA: Jordan Valley Authority

KDI: Kounkuey Design Initiative

LRWHF: Lanka Rain Water Harvesting Forum

LYDEC: Lyonnaise de Eaux Casablanca

MARI: Modern Architects for Rural India

MENA: Middle East and North Africa

MUT: multi-utility tunnel

NDRC: National Development and Reform Committee (China)

NGO: Nongovernment organization

NHPC: National Hydro Power Corporation

NMC: Nashik Municipal Corporation

OECD: Organisation for Economic Cooperation and Development

OECF: Overseas Economic Cooperation Fund

OMC: Omnigrid Micropower Co., Pvt., Ltd.

OPEC: Organization of Petroleum Exporting Countries

PADE: Programme d'Assainissement de Diokoul et quartiers Environnants (Diokoul and Surrounding Districts Sanitation Scheme)

PIDA: Programme for Infrastructure Development in Africa

PPP: public-private partnership

PSH: pumped storage hydro

PUB: Public Utilities Board (Singapore)

PV: photovoltaic

RIBA: Royal Institute of British Architects

RWP: Reclaimed Water Project

SEPA: State Environmental Protection Administration (China)

SLU: Superintendency of Public Cleansing

SMART system: Stormwater Management and Road Tunnel system

SSNNL: Sardar Sarovar Narmada Nigam Ltd.

SWG: seawater greenhouse

SWH: solar water heater

TEDA: Tianjin Economic-Technological Development Area

ULI: Urban Land Institute

UN: United Nations

UNDP: United Nations Development Programme

USAID: United States Agency for International Development

VHW: Village Health Works

VNRC: Vietnam Red Cross

WWF: World Wildlife Fund

Glossary

Absolute scarcity A condition that exists when there is not enough of a resource in existence to satisfy existing demand for it.

Adaptation Adjustments in human and/or natural systems in response to actual or expected climate stimuli or their effects that moderate harm or exploit beneficial opportunities.

Appropriate technology Technology that is suitable to the social and economic conditions of the geographic area in which it is to be applied, is environmentally sound, and promotes self-sufficiency on the part of those using it.

Aquaponics A system of aquaculture in which the waste produced by farmed fish or other aquatic animals supplies nutrients for plants grown hydroponically, which in turn function to purify the water.

Bankable (infrastructure) project A project or proposal that has sufficient collateral and future cashflow, and a high enough probability of success, for institutional lenders to be willing to provide financing.

Base load The constant or permanent load on a power supply.

Biocide (biocidal) A biocide is defined in European legislation as a chemical substance or microorganism intended to destroy, deter, render harmless, or exert a controlling effect on any harmful organism by chemical or biological means.

Biodigestion (anaerobic digestion) Processes by which microorganisms break down biodegradable material *in the absence of oxygen* to manage waste and/or produce fuels. The process initially relies on bacterial hydrolysis of the waste inputs. Insoluble organic polymers (e.g., carbohydrates) are broken down to soluble derivatives made available to other bacteria. These acidogenic bacteria convert sugars and amino acids into carbon dioxide, hydrogen, ammonia, and organics. Lastly, methanogens (microorganisms that produce methane in anoxic conditions) convert these products to methane and carbon dioxide.

Biodiversity   The variety of life in the world or in a particular habitat or ecosystem.

Biogas   A fuel derived from the decay of organic matter as the mixture of methane and carbon dioxide produced by the bacterial decomposition of sewage, manure, garbage, or plant crops.

Biosolids   Organic matter recycled from sewage, especially for use in agriculture.

Blackwater   Wastewater containing bodily or other biological wastes, as from toilets, dishwashers, or kitchen drains.

Bund   Term for a dam used in India and Southeast Asia.

Cascading resource use   Gaining multiple uses from a single resource (for example, water or energy) by utilizing different grades for different applications rather than using the highest-quality energy or resource for all uses.

Casuarina (trees)   Evergreen shrubs and trees native to Australia, the Indian subcontinent, Southeast Asia, and islands of the western Pacific Ocean.

Certified Emission Reduction (CER) credits   A certificate issued every time the United Nations prevents one metric ton of CO_2 equivalent being emitted through carbon projects registered with the CDM.

C40 Climate Leadership Group   A network of the world's megacities committed to addressing climate change.

Circular economy   An alternative to a traditional linear economy (make, use, dispose) in which we keep resources in use for as long as possible, extract the maximum value from them while in use, and then recover and regenerate products and materials at the end of each service life.

Clean Development Mechanism (CDM)   Established under the original Kyoto Protocol of 1997, developed countries that have made emission-reduction commitments can implement emission-reduction projects in developing countries. Such projects can earn saleable certified emission reduction (CER) credits that can be counted toward meeting Kyoto targets.

Climate departure   The term describing the onset of a regime where once stochastic flooding and other damaging meteorological events become the new normal.

Closed-loop system   A system in which waste streams become inputs to secondary or tertiary processes.

Cogeneration   The generation of electricity and other energy jointly, especially the utilization of the steam left over from electricity to produce heat.

Colocation   The placement of several entities in a single location.

Commensalism   An association between two organisms in which one benefits and the other derives neither benefit nor harm.

Cost-benefit analysis   A systematic approach to estimating the strengths and weaknesses of alternatives regarding transactions, activities, or functional requirements for a business.

Credit enhancement   An improvement of the credit profile of a structured financial transaction or the methods used to improve the credit profiles of such transactions.

Debt overhang   Significant debt from prior administrations.

Departmentalized system   A system planned, designed, developed, and implemented by specialists who have little or no contact with their counterparts in other agencies or fields.

Desalination   A process to obtain fresh water suitable for drinking or irrigation by removing minerals from saline or seawater, particularly salts (sodium chloride). Processes include distillation, essentially boiling water at less than atmospheric pressure, and membrane processes using semipermeable films to filter out dissolved material or fine solids.

Dissolved oxygen   The microscopic bubbles of gaseous oxygen mixed in water and available to aquatic organisms for respiration.

Distributed or decentralized infrastructure   A shift away from large, centralized infrastructure facilities toward smaller district-, neighborhood-, or home-level alternatives to meet local needs.

Ebullition   The sudden release of bubbles of methane into the air as a result of methane building up over time in the soil, forming pockets of methane gas. As these pockets grow in size, pressure builds up until the bubble "pops."

Economies of scale   A proportionate saving in costs resulting from an increase in the level of production.

Ecosystem services   The multifaceted benefits provided by ecosystems and their components (water, soil, nutrients, organisms), including clean air, water, food, and materials that serve human well-being and health. These break down into four categories:

• **Regulating services** include air-quality regulation, climate regulation, water purification, erosion control, pollination, and disease regulation.
• **Provisioning services** include food, fiber, fuel, biochemicals, pharmaceuticals, and natural medicines obtained through ecosystem activities.
• **Supporting services** include photosynthesis, primary production, nutrient cycling, soil formation, and water cycling.
• **Cultural services** include nonmaterial benefits from ecosystems, such as spiritual enrichment, cognitive development, recreation, and aesthetic experiences.

Energy cascading   The transfer of the remaining energy in steam or liquids to provide energy (as heat, cooling, or pressure) for another process to reduce overall entropy of a given system.

Energy, water, and food nexus   The inextricable links among these three sectors: water essential to produce agricultural goods; energy required to produce and distribute water and food; and water necessary to provide power plant cooling. Balancing trade-offs is central to jointly ensuring security of water, energy, and food.

Evapotranspiration   The process by which water is transferred from the land to the atmosphere by evaporation from the soil and transpiration by trees and plants (water absorbed through roots and given off as water vapor through pores in their leaves).

Feasibility study   An assessment of the practicality of a proposed plan or method.

Forestry certification program   Processes through which forest owners, managers, and the whole wood chain of custody activities are certified through independent parties that the production of wood is in compliance with certain sustainable standards.

Gabion   A basket or cage filled with earth or rocks used especially for fortification or abutment; for example, to support pieces of ground or control a flow of water.

Global positioning system (GPS)   A space-based navigation system that provides location and time information in all weather conditions anywhere on or near the earth where there is an unobstructed line of sight to four or more GPS satellites.

Greenfield site   Land (as a potential industrial site) not previously developed or polluted.

Greywater   All wastewater generated in households or office buildings from streams without fecal contamination; that is, all streams except for the wastewater from toilets. Sources of greywater include, for example, sinks, showers, baths, clothes washing machines, or dishwashers.

G20 countries   Finance ministers and central bank governors of 19 countries: Argentina, Australia, Brazil, Canada, China, France, Germany, India, Indonesia, Italy, Japan, Mexico, Russia, Saudi Arabia, South Africa, South Korea, Turkey, the United Kingdom, and the United States.

Hydroponic farming   A (high-tech) growing method where vegetables are grown without soil. The roots of plants are fed a nutrient solution or grown in media such as perlite or gravel.

Hypolimnion   The methane-rich lower layer of water in a stratified lake, typically cooler than the water above and relatively stagnant.

Industrial ecology   The study of energy and material flows in industrial systems in order to identify opportunities to reduce the consumption of energy and resources while minimizing pollution and waste.

Industrial symbiosis   A method used to determine how efficiencies could be obtained through links between diverse producers.

Infrastructural ecology   A conceptual model for addressing both economic and social values with regard to the development of infrastructure. It capitalizes on synergies from interconnected flows of energy, water, and matter to create multifunctional or networked structures.

Infrastructural silos   The habit of seeing each component of an infrastructural project as a discrete object instead of as a part of a unified system. This obscures the primacy of interrelationships in complex systems and precludes our gaining a grasp of dynamic, complex system behavior.

Internalized externalities   The act of making a change in private costs or benefits to make them equal to social costs or benefits, accounting for true monetary costs of development choices.

Kyoto Protocol   An international treaty among industrialized nations that sets mandatory limits on greenhouse gas emissions.

Legacy systems   In the context of infrastructure systems, older methods or technologies, often implying that the systems are out of date or in need of replacement.

Littoral zones   The area near a sea, river, or lakeshore where sunlight penetrates all the way to the sediment and allows growth of aquatic plants.

Mariculture   Specialized aquaculture around cultivation of marine organisms for food (or other products) in open or enclosed ocean areas or in other bodies filled with seawater.

Metabolic flows   A series of steps that help convert energy, material, water, and waste into different, more readily usable materials.

Microgrid   A group of interconnected loads and distributed energy resources within clearly defined electrical boundaries that acts as a single controllable entity with respect to the grid. A microgrid can connect and disconnect from the grid to enable it to operate in both grid-connected or island mode.

Mission creep   A gradual shift in objectives or expansion of a project beyond its original goals.

Multilateral development bank (MDB)   An institution created by a group of countries that provides financing and professional advice for the purpose of economic and social development. MDBs have large memberships, including both developed donor countries and developing borrower countries.

Natural capital   The term is a critique of traditional "industrial capitalism" that says that the traditional system of capitalism "does not fully conform to its own accounting principles. It liquidates its capital and calls it income." The world's stock of natural assets, which include geology, soil, air, water, and all living things.

Oleaginous waste   Waste rich in, covered with, or producing oil.

100 Resilient Cities   An initiative pioneered by the Rockefeller Foundation to help more cities build resilience to the physical, social, and economic challenges that are a growing part of life in the twenty-first century.

One-hundred-year flood   A flood that statistically has a 1 percent chance of occurring in any given year.

Permaculture   A system of agricultural and social design principles centered on simulating or directly utilizing the patterns and features observed in natural ecosystems.

Polytunnel   Also known as a polyhouse, hoop greenhouse or hoophouse, or high tunnel, it is a tunnel made of polyethylene and usually semicircular, square, or elongated in shape.

Positive feedback loop   A process whereby one action has a positive result that enhances or amplifies changes in another activity. This often results in moving a system away from its equilibrium state and making it more unstable.

Poverty trap   A mechanism that makes it very difficult for people to escape poverty. It is created when an economic system requires significant amounts of various forms of capital in order to earn enough to escape poverty.

Predevelopment expenses   Development costs a business or public entity incurs from researching, growing, or introducing a new service.

Principal city   The largest incorporated place in a core-based statistical area with a population of at least 10,000, or if no such incorporated place exists, the largest incorporated or census-designated place.

Pro forma   Latin term meaning "for the sake of form." Assumed, forecasted, or informal information presented in advance of a transaction to satisfy minimum requirements.

Public-private partnerships   A contractual arrangement between a public agency (federal, state, or local) and a private sector entity whereby the skills and assets of each sector (public and private) are shared in delivering a service or facility for the use of the general public.

Pumped storage hydro   Projects that store and generate energy by moving water between two reservoirs at different elevations. At times of low electricity demand, excess energy is used to pump water to an upper reservoir.

Regenerative development   A system of technologies and strategies for generating the patterned whole system understanding of a place and for developing the strategic systemic thinking capacities and the stakeholder engagement/commitment required to ensure regenerative design processes to achieve maximum systemic leverage and support that is self-organizing and self-evolving.

Restorative design   Sometimes called restorative environmental design, it is a design system that returns degraded or polluted land back to a state of acceptable health through human intervention.

Reverse osmosis (RO)   A water purification technology that uses a semipermeable membrane to remove larger particles from drinking water.

Risk-mitigation assessment   A review conducted to reduce adverse effects or exposure to a risk and/or the likelihood of its occurrence.

Self-organized (self-regulated) process   A process where some form of overall order or coordination arises out of the local interactions between smaller component parts of an initially disordered system.

Semiarid   Place or climate zone that is partially arid or semidry and has less than 20 inches of rain each year.

Soft (water) path   A more integrated and effective alternative to supply-side water resource management. Whereas "hard path" infrastructure relies almost exclusively on centralized infrastructure, the "soft path" also considers decentralized facilities, efficient technologies, reliance on ecological processes, and natural and human capital.

Source to sink   Simple linear flows from resource sources (farms, mines, forests, watersheds, oilfields, etc.) to sinks (air, water, land) that deplete global sources and overload or pollute global sinks.

Subsistence-level farming   Self-sufficiency farming in which farmers focus on growing enough food to feed themselves and their families.

Synergies   The interaction or cooperation of two or more organizations, substances, or other agents to produce a combined effect greater than their separate effects.

Systems thinking   Also known as whole-systems thinking, it is a framework for seeing interrelationships rather than things, understanding phenomena in terms of the whole rather than parts, and developing solutions to problems by considering the interconnections between systems so that human and environmental problems can be addressed simultaneously.

Transit-oriented development   A mixed-use residential and commercial area designed to maximize access to public transport. It often incorporates features to encourage transit ridership.

United States Foreign Corrupt Practices Act (FCPA)   Criminalizes bribery of foreign officials in exchange for business opportunities, levying major fines on companies or individuals found to be engaging in such practices.

Urban heat island   The presence, in a city or metropolitan area, of temperatures that are significantly warmer than surrounding rural areas as a result of human activities. This includes the prevalence of dark surfaces (roofs and paving) that retain rather than reflect infrared light and the accumulation of waste heat (most notably air conditioners and internal combustion engines).

Urbanization   The process by which towns and cities are formed and become larger as more and more people begin living and working in central areas.

Urban metabolism   Similar to a living organism's metabolism, it involves resource consumption and waste generation by an urban area. In a linear metabolism, resources are extracted external to the city, and wastes are removed there. Circular metabolisms reuse wastes repeatedly within the urban boundary to maximize the value derived from resources.

User fees   A fee, charge, or tax paid to the utility operator by a customer as a necessary condition for utility services.

Recommended Readings

Chapter 1

Alonso, Jose Antonio, Andrea Cornia Giovanni, and Rob Vos, eds. *Alternative Development Strategies for the Post-2015 Era*. London: Bloomsbury, 2014.

Bhattacharya, Amar, Mattia Romani, and Nicholas Stern. "Infrastructure for Development: Meeting the Challenge." Centre for Climate Change Economics and Policy, Grantham Research Institute on Climate Change and the Environment, and the Intergovernmental Group of Twenty Four, Policy Brief, June 2012, http://gggi.org/wp-content/uploads/2013/01/Policy+Brief_2013_pb1.pdf.

Brown, Hillary. *Next-Generation Infrastructure*: *Principles for Post-industrial Public Works*. Washington, DC: Island Press, 2014.

Collier, Paul. *The Bottom Billion: Why the Poorest Countries Are Failing and What Can Be Done about It*. New York: Oxford University Press, 2007.

———. *The Plundered Planet: Why We Must—And How We Can—Manage Nature for Global Prosperity*. New York: Oxford University Press, 2011.

Escobar, Arturo. *Encountering Development: The Making and Unmaking of the Third World*. Princeton, NJ: Princeton University Press, 2011.

The Global Commission on the Climate and the Economy. *Better Growth/Better Climate: The New Climate Economy Report*. Washington, DC: World Resources Institute, 2014.

Miller, Marian. *The Third World in Global Environmental Politics*. London: Rienner, 1995.

Pinderhughes. Raquel. *Alternative Urban Futures: Planning for Sustainable Development in Cities throughout the World*. Lanham, MD: Rowman and Littlefield, 2006.

United Nations, Department of Economic and Social Affairs, Population Division. *World Urbanization Prospects: The 2014 Revision, Highlights*. ST/ESA/SER.A/352. Geneva: United Nations, 2014.

The World Bank: The International Bank for Reconstruction and Development. *Infrastructure at the Crossroads: Lessons from 20 Years of World Bank Experience*. Washington, DC: World Bank, 2006.

Chapter 2

Girardet, Herbert. *The Gaia Atlas of Cities: New Directions for Sustainable Urban Living*. London: Gaia Books, 1992.

Spretnak, Charlene. *Relational Reality: New Discoveries of Interrelatedness That Are Transforming the Modern World*. Topsham, ME: Green Horizon Books/Green Horizon Foundation, 2011.

Chapter 3

Daily, Gretchen, ed. *Nature's Services: Societal Dependence on Natural Ecosystems*, fourth edition. Washington, DC: Island Press, 1997.

Gleick, Peter H., and associates. *The World's Water: The Biennial Report on Freshwater Resources*, volume 8. Washington, DC: Island Press, 2014.

Hardoy, Jorge E., Diana Mitlin, and David Satterthwaite. *Environmental Problems in an Urbanizing World: Finding Solutions in Cities in Africa, Asia and Latin America*. New York: Earthscan Publications, 2001.

Pittock, Jamie, ed. "Water for Life: Lessons for Climate Change Adaptation from Better Management of Rivers for People and Nature." Gland, Switzerland: World Wildlife Fund, 2008.

Solomon, Steven. *Water: The Epic Struggle for Wealth, Power, and Civilization*. New York: Harper-Collins, 2010.

Chapter 4

Brown, Lester R. *The Great Transition: Shifting from Fossil Fuels to Solar and Wind Energy*. New York: Earth Policy Institute and W. W. Norton, 2015.

———. *Plan B 4.0: Mobilizing to Save Civilization*. New York: Earth Policy Institute and W. W. Norton, 2009.

Energy Storage. "Packing Some Power." Technological Quarterly, *The Economist*, March 3, 2011, http://www.economist.com/node/21548495?frsc=dg%7Ca.

Goldthau, Andreas. "Rethinking the Governance of Energy Infrastructure: Scale, Decentralization and Polycentrism." *Energy Research and Social Science* 1 (2014): 134–140.

Ramos, F. M., L. W. W. Bambace, I. B. T. Lima, R. R. Rosa, E. A. Mazzi, and P. M. Fernside. "Methane Stocks in Tropical Hydropower Reservoirs as a Potential Energy Source." *Climatic Change* 93 (2009): 1–13.

Renewable Energy Policy Network for the 21st Century (REN21). *Renewables 2010 Global Status Report*. Paris: REN21 Secretariat, 2010.

Schuster-Wallace, C. J., C. Wild, and C. Metcalfe. "Valuing Human Waste as an Energy Resource: A Research Brief Assessing the Global Wealth in Waste." United Nations University Institute for Water, Environment, and Health (UNU-INWEH), 2015.

Srinivasan, R. "In Areas Where Grid Power and Diesel Supply Is [*sic*] Erratic, Solar Power Has a Strong Business Case." *Power Watch India*, June 4, 2015, http://powerwatchindia.com/in-areas-where-grid-power-and-diesel-supply-is-erratic-solar-power-has-a-strong-business-case/.

Terminski, Bogumil. *Development-Induced Displacement and Resettlement: Theoretical Frames and Current Challenges*. Geneva: University of Geneva, 2013.

United Nations. "The Millennium Development Goals Report." Geneva: United Nations, 2015.

Chapter 5

Baker, Judy L. *Climate Change, Disaster Risk and the Urban Poor: Cities Building Resilience for a Changing World*. Washington, DC: World Bank Publications, 2012. http://elibrary.worldbank.org/doi/abs/10.1596/978-0-8213-8845-7.

Bird, Jeremy. "Water-Food-Energy Nexus." In *Setting and Achieving Water-Related Sustainable Development Goals*, edited by Julie Van Der Bliek, Peter McCormick, and James Clarke, chapter 2. Colombo, Sri Lanka: International Water Management Institute, 2014. http://www.iwmi.cgiar.org/Publications/Books/PDF/setting_and_achieving_water-related_sustainable_development_goals-chapter-2-water_food_energy_nexus.pdf.

Burton, I., E. Diringer, and J. Smith. "Adaptation to Climate Change: International Policy Options." Washington, DC: Pew Center on Global Climate Change, 2006. http://www.c2es.org/docUploads/PEW_Adaptation.pdf .

Fletcher, Roland. "Low-Density, Agrarian-Based Urbanism: A Comparative View." *Insights, University of Durham* 2, no. 4 (2009): 2–19.

The Global Commission on the Climate and the Economy. *Better Growth/ Better Climate: The New Climate Economy Report*. Washington, DC: World Resources Institute, 2014.

Hunt, Constance Elizabeth. *Thirsty Planet: Strategies for Sustainable Water Management*. Chicago: Zed Books and University of Chicago Press, 2004.

Hussain, Intizar, Liqa Raschid, Munir A. Hanjra, Fuard Mrikar, and Wim van der Hoek. "Wastewater Use in Agriculture: Review of Impacts and Methodological Issues in Valuing Impacts." International Water Management Institute Working Paper 37, Colombo, Sri Lanka, International Water Management Institute, 2002.

Savory, Allan. "Holistic Resource Management: A Conceptual Framework for Ecologically Sound Economic Modeling." *Ecological Economics* 3 (1991): 181–191.

UNEP-WCMC. "In the Front Line: Shoreline Protection and other Ecosystem Services from Mangroves and Coral Reefs," 2006, 1–33, http://www.unep.org/pdf/infrontline_06.pdf.

UN Water. "Climate Change Adaptation: The Pivotal Role of Water," Policy Brief, 1–18, http://www.unwater.org/downloads/unw_ccpol_web.pdf.

U.S. Environmental Protection Agency. "Reducing Urban Heat Islands: Compendium of Strategies," https://www.epa.gov/heatisland/resources/pdf/BasicsCompendium.pdf.

The World Bank. *The Cost to Developing Countries of Adapting to Climate Change: New Methods and Estimates*. Washington, DC: World Bank, 2010.

———. *World Development Report: Development and Climate Change*. Washington, DC: World Bank, 2010.

Chapter 6

Arnstein, Sherry R. "A Ladder of Citizen Participation." *American Institute of Planning Journal* 35, no. 4 (July 1969): 216–224.

Herbertson, Kirk, Athena Ballesteros, Robert Goodland, and Isabel Munilla. *Breaking Ground: Engaging Communities in Extractive and Infrastructure Projects*. Washington, DC: World Resources Institute, 2009. http://pdf.wri.org/breaking_ground_engaging_communities.pdf.

Mansuri, Ghazal, and Vajayendra Rao. *Localizing Development: Does Participation Work?* Washington, DC: World Bank, 2013.

Schnurr, Jamie, and Susan Holtz. *The Cornerstone of Development: Integrating Environmental, Social, and Economic Policies*. Boca Raton, FL: CRC Press, 1998.

Chapter 7

Hunter, David. "Using the World Bank Inspection Panel to Defend the Interests of Project-Affected People." *Chicago Journal of International Law* 4, no. 1 (2003): 201–211.

Sacks, Laura, and Erik Van Voorthuysen. "Energising Innovation: The Role of Global Innovation Alliances in Addressing Key Energy Challenges." *Global Voices Journal*, August 15, 2015, https://globalvoices.org.au/journal/laura-sacks-y20research.

Chapter 8

Brugmann, Jeb, with Katherine Brekke and Lucy Price. "Operationalizing the Urban NEXUS: Towards Resource-Efficient and Integrated Cities and Metropolitan Regions." Bonn: Deutsche Gesellschaft für Internationale Zusammenarbeit GmbH and ICLEI, 2014. http://resilient-cities.iclei.org/fileadmin/sites/resilient-cities/files/Full_papers/Urban_NEXUS_Publication_ICLEI-GIZ_2014_web.pdf.

United Nations World Water Assessment Programme (WWAP). *The United Nations World Water Development Report 2014: Water and Energy*. Paris: UNESCO, 2014.

Index

Page numbers for figures are in italics.

100 Resilient Cities, 179, 184

Abengoa, 141
Adeyemi, Kunlé, 126
Africa Centre for Holistic Management
  (ACHM), 139
agriculture. *See also* polyculture (agriculture
  and aquaculture)
 Bangladesh floating agricultural system, 123
 improvements through water management,
  58–59
 Nashik Urban NEXUS program, 158–59
Agroenergy Condominium for Family
  Agriculture, 102
agroinfrastructure, 25
Ahmedabad Heat Action Plan, 130
Algae Biomass Consortium, Department
  of Energy, 144
Anna Watta Kai (Well-being for the Future)
  Foundation, 156
Anse-Rouge microindustrial park, 208–11,
  280n28
Anse-Rouge Seawater Greenhouse (SWG), 209
aquaculture, 23, 41, 69, 71, 120, 136, 162
artificial floating islands (AFIs) in freshwater
  lakes, China, 124

Association of Collectors of Paper, Cardboard
  and Recyclable Material (ASMARE), 167
Aztecs. *See* Chinampas in pre-Columbian
  Mesoamerica

Belo Horizonte greenhouse gas mitigation
  projects, 99–101
Belo Horizonte's waste pickers, 166
Berry, Simon, 30
Berry, Wendell, 19
Betim Reference Center in Renewable
  Energy, 93
Betim solar thermal energy project, 92–93
biogas recovery, 73–74. *See also* wastewater
  treatment
 Agroenergy Condominium for Family
  Agriculture, 102–3
 Belo Horizonte landfill methane recovery,
  99–101
 biogas-producing waste facilities, 97–98,
  256n77
 Kigutu wastewater and solid waste
  treatment, 96
 Lille, treated biogas for municipal buses,
  259n124
 Metrogas Town Gas Plant (Santiago), 74

biogas recovery (cont.)
 Nashik Urban NEXUS program, 160
 Nepal Biogas Support Program, 201
Biogas Support Program, 201
biomass
 biofuels from bagasse, 92
 DOE Algae Biomass Consortium, 144
 energy from biodigested agricultural residue,
  102
 Oujé-Bougoumou biomass-fueled district
  heating system, 94
 residual and renewable, 92
 sustainability, 83
biopores. See Bogor City "biopores"
Bogo Institute of Agriculture (Indonesia), 64
Bogor City "biopores," 63–66
Brata, Kamir R., 64
Brown, Hillary, 5
building codes, modernization
 International Building Code (IBC), 189
 International Standards Organization (ISO),
  189
 Royal Institute of British Architects (RIBA),
  188
Burundi, urbanization of, 7

Canary Islands, electric vehicle goal, 90
Canary Islands Institute of Technology, 89
Caracol Industrial Park, 2–4, 207–8, 236n12
carbon emissions, 113–14. See also CO_2
  abatement
 aerial cable cars, 108–10
 car ownership or low-carbon transportation,
  107–8, 114, 202, 204, 259n124
 developing nations and, 78–80, 252n7
 fuel and energy generation costs, 80–81
 low-carbon and renewable energy,
  199–200
Cardinal, Douglas, 94
Casablanca Temporary Lower Cost Mini-Grid
  Electrification program, 157–58
Chinampas in pre-Columbian Mesoamerica,
  22–25

circular economies, 40–41
 in China, 46
 Tianjin Economic-Technological
  Development Area (TEDA), 43, 44, 244n82
climate change, 203. See also floating
  platforms; sea-level rise and coastal
  resilience
 eutrophication in freshwater lakes, China, 124
 infrastructure adaptations, 115–16, 144
 mangrove buffers role, 120, 161–62
 rising temperature and heat islands, 129–31
 World Bank study on cost of adaptation, 116
closed-loop systems. See infrastructural
  ecology; Seawater Greenhouse (SWG)
CO_2 abatement, 78, 79, 94
 Clean Development Mechanism (CDM),
  177–78
 Humbo Assisted Natural Regeneration
  Project, 66–67
 Ibadan biomethane recovery project, 98
 Jepírachi project, pilot windfarm, 154
 Mekong Delta mangrove restoration, 122
 Metrogas Town Gas Plant (Santiago), 74
 Tianjin Economic-Technological
  Development Area (TEDA), 45
 Trung Son hydropower project, 38
ColaLife, 31, 241n36
 distribution with Coca Cola, 30
Collier, Paul, 29
colocation, 29, 238n41. See also community
  participation; "multi-utility tunnel" (MUT)
 agriculture and desalination, 25
 Sardar Sarovar Narmada Nigam Ltd, 183
 Treatment Center for Solid Waste (CTRS),
  100
 Zambia ColaLife distribution, 30
commensalism, 29
Commune Project Coordinating Committees
  (CPCCs), 152–53
community participation, 4, 55, 60, 169, 174
 Belo Horizonte's waste pickers, 166–68
 benefits of, 150
 Bogor City "biopores," 66

Casablanca Temporary Lower Cost Mini-Grid
  Electrification program, 157–58
Community-Based Mangrove Management
  Board (CMMB), 162–64
Cultivando Agua Boa—Cultivating Good
  Water, 104
decentralization, 151–53, 206–7
Diokoul and Surrounding Districts Sanitation
  Scheme, 164–66
East Kolkata Wetlands (EKW), 72
Enkanini iShack, 81–82
Humbo Assisted Natural Regeneration
  Project, 67
India, Manor subbasin, 58–60
Ixtlán de Juárez Communal Forest, 170–72
Jepírachi project with the Wayuu
  community, 154–56
Kenya Kingapop Wind farm local
  resistance, 185
Las Gaviotas, Colombia, 33
Lufumbu community water management
  project, 172–73
Makoko/Iwaya Waterfront Regeneration
  Plan, 125–27
Nashik Urban NEXUS program, 158, 160–61
Nashik's Sustainable Urban Habitat Action
  Plan, 161
Nepal microhydropower projects, 84
Peruvian bridge rebuilding ritual, 147–49
Sri Lanka rainwater harvesting, 55
stakeholder engagement, 185–86
Community Water Supply and Sanitation
  Project (CWSSP), 54
Community-Based Mangrove Management
  Board (CMMB), 162–64
cool roofs. See climate change, rising
  temperature and heat islands
Curitiba, green infrastructure in, 61–63
Curitiba Municipal Decree for Riverside Areas
  Preservation, 61

Da Loc mangrove restoration, 161–62
Da Nang "environmental city," 108

Delhi Street Design Guidelines, tree planting,
  130
desalination, 205
  El Hierro, Canary Islands, 88
  indirect solar desalination, 140–41
  Samra wastewater plant, 39
Deutsche Gesellschaft für Technische
  Zusammenarbeit GmbH (GTZ), 137
Dimbangombe grassland pilot project, 139–40
Diokoul and Surrounding Districts Sanitation
  Scheme, 164–66

EarthSpark International, 34–35
  SparkMeter (smart meter), 34
East Kolkata Wetlands (EKW), 69–72
ecoindustrial parks (EIP), 46
  Anse-Rouge, Haiti, 208–9
  in China, 43, 45–46, 177
  Kalundborg, Denmark, 13
El Hierro, Canary Islands wind and pumped
  storage hydro, 89
Empresas Públicas de Medellín (EPM), 154, 156
energy cascades, 40, 44, 45
energy generation. See hydroelectric energy
  generation; solar energy; wastewater energy
  generation; wind energy generation
engineered water infrastructure, 55
Enkanini iShack, 81–82
environmental and social impacts, 11, 46, 56.
  See also Caracol Industrial Park
  Curitiba, green infrastructure in, 62
  pollution and waterborne disease in
  Indonesia, 63

Federal Energy Regulatory Commission,
  United States, 91
Fergana Valley water banking, 133–34, 136
floating platforms, 23
  artificial floating islands, 124
  floating agricultural system, 123
  floating solar power station, 37
  Waterbuurt, floating community on
  Lake IJ, 124

flooding and flood control, 60, 116, 248n59
  Bangladesh, 122
  Kibera, Nairobi integrated flood-mitigation
    process, 128, *129*
  Kolkata, India, 72
  Makoko, Lagos, 125
  Red River Delta, Viet Nam, 121, 161
Foreign Corrupt Practices Act (FCPA), 191
forest management, 60, 66–67, 169–72
  silviculture, 171, 275n96
funding. *See also* politics of infrastructure
  projects
  Clean Development Mechanism (CDM), 177
  credit enhancement, 179
  multiple revenue streams, 177–78
  predevelopment funding, 178–79
  public-private partnership laws (PPP), 180–81
  short and long term, 176
funding, external, 8. *See also* United Nations;
  World Bank
  African Development Bank, 185
  Asian Development Bank (ADB), 19, 108,
    131, 179, 185
  Asian Infrastructure Investment Bank (AIIB),
    179
  Bill and Melinda Gates Foundation, 82
  Canadian Host Country Participation Fund,
    165
  CARE, 122, 124, 162–64, 207
  Catholic Church, 167
  Clinton Global Initiative, 185
  ENDA-Third World/RUP, 164–66
  Fundación Chile, 141
  German Federal Ministry for Economic
    Collaboration and Development, 137
  German Society for International
    Cooperation (GIZ), 158–59
  Global Network for Environment and
    Economic Development Research, 98
  Inter-American Development Bank (IDB),
    2, 185, 236n12
  International Institute for Environmental
    Development (IIED), 124

International Monetary Fund (IMF), 31
  Modern Architects for Rural India (MARI), 58
  Overseas Economic Cooperation Fund
    (OECF), 19
  Rockefeller Foundation, 179, 185
  Urban Climate Change Resilience
    Trust Fund, 179
  USAID, 35, 39, 53, 54, 56
  Viet Nam Red Cross, 121, 161
  World Vision Australia, 66
  World Vision Ethiopia, 66, 67
  World Wildlife Fund, 58, 59, 170
funding, government
  cross agency partnerships, 183
  Dutch and Indonesian Jakarta land
    reclamation project, 119–20
  Egypt, PPP Central Unit, 181
  GEMIG, Brazil state electrical utility, 92
  German Federal Ministry for Economic
    Collaboration and Development, 158
  Itaipu Binacional (IB) by Brazilian
    government, 102
  National Hydro Power Corporation (NHPC),
    Japan, 37
  Nepal Electricity Authority, 84
  Postal Solecshaw for mail delivery in India, 111
  Rufisque Local Authority (Senegal), 165
  Spain and EU for El Hierro, 89–90
  Telangana, India, 59

Girardet, Herbert, *40*
Global Population Projections, 1950–2050
  (United Nations), *6*
grassland ecology, 138–40
Greater Angkor, water management, 116–*17*
green infrastructure, definition of, 60
greenhouse gas emissions, 13, 18, *78–79*, 92,
  102, 252n4. *See also* methane recovery
  submerged vegetation methane generation, 38

Hammarby Sjöstad, Stockholm, 14–16, 45–46
Humbo Assisted Natural Regeneration Project,
  66–68

hydroelectric energy generation
  dam for energy generation and irrigation,
    37–38
  Itaipu Binacional (IB), 102
  Liberia mini hydropower facility, 38–39
  Mein River "run-of-the-river" project, 38–39
  methane recovery, 105–6
  Nepal, microhydropower, 83–84
  pumped storage hydro (PSH), 88–89
  Renewable hydro and aerodynamic energy, 77
  Trung Son hydropower project, 38

Ibadan biomethane recovery project, 98–99
Ikaria Island combined wind and pumped
    storage hydro, 91
industrial ecology
  definition, 5
  "Hammarby Model," 15
  industrial symbiosis, 12–14
  objectives, 16–17
industrial symbiosis, 12–13, 46, 238n42
Industry and Environment (United Nations), 43
infrastructural ecology, *40*
  advantages of, 11–12
  in China, 46
  definition, 5
infrastructure-sharing strategies, 193–94,278n1
Institute for Urban Planning Research (IUPR), 63
integrated planning, 7, 10–11, 80, 97, 151,
    194–95
integrated technical expertise, 186–88
Intergovernmental Panel on Climate Change
    (IPCC), 115
International Council for Local Environmental
    Initiatives (ICLEI), 92, 112, 158–59
International Energy Agency, 106
International Fund for Agricultural
    Development (IFAD), 152
International Water Management Institute
    (IWMI), 53, 136
irrigation, 35
  Chinampas in pre-Columbian Mesoamerica,
    22–25

East Kolkata Wetlands (EKW), 71
  raised field cultivation and lake-based
    development, 22
  solar arrays over irrigation canals (India),
    36–37
  Soviet era Aral Sea disappearance, 266n94
  Sri Lanka ancient irrigation system, 48–51
  treated wastewater quality guidelines,
    137–38
Itaipu Binacional (IB), 102
Itaipu Technological Park (ITP), 104–5
Ixtlán de Juárez Communal Forest, 169–72,
    207

Jakarta, flooding, 118–19
Jakarta, seawall and land reclamation project,
    119
Jakarta, traffic and flood control tunnels, 119
Jakarta Coastal Defense Strategy, 119
Jamuna Bridge, 19–20, 178, 183, 186
Jawaharlal Nehru National Solar Mission, 35
Jepírachi project, pilot wind farm, 153–57
Jordan Valley Reclaimed Water Project (RWP),
    137–38

Kagoshima Nanatsujima Mega Solar Power
    Plant, 37
Kalundborg, Denmark, 13–14
Khaju Bridge, Isfahan, 21–22
Khmer empire hydraulic network, 116–18
Kibera, Nairobi flooding, 127–28, *129*
Kigutu wastewater and solid waste treatment,
    95–99
Kinangop Wind Farm, 180, 185
Kounkuey Design Initiative (KDI), 128
Kuala Lumpur SMART system, 27–29, 183
Kyoto Protocol, 177
  Clean Development Mechanism (CDM), 100

Lagarde, Christine, 31
Lake Kivu, methane potential, 106–7
Lake Sihwa tidal estuary project, 39
Lake Texcoco, island-city Aztec capital, 22–23

landfill methane capture. *See* biogas recovery

Lanka Rainwater Harvesting Forum
  (LRWHF), 55

Las Gaviotas seesaw pump, 32–33

legal frameworks, 190
  U.S. Foreign Corrupt Practices Act (FCPA), 191

Lerner, Jaime, 63

Les Anglais, excess energy sold, 34

Lima water management by SEDEPAL,
  133–34, *135*

Lluta Valley solar-powered desalination plan,
  141

low-carbon infrastructure. *See* carbon
  emissions

Lufumbu community water management
  project, 172–73

Lugari, Paulo, 32

Lyonnaise des Eaux Casablanca (LYDEC),
  157–58

Makoko, Lagos schoolhouse project, 125

Makoko/Iwaya Waterfront Regeneration Plan,
  125–27

Maner subbasin, 55–59

Martelly, Michel, 2

Medellin Metro-Cable Line-K, 108

Mekong Delta mangrove restoration, 120–22

methane recovery, 105–6, 201, 251nn112–113
  Belo Horizonte landfill methane recovery,
  99–101
  Ibadan biomethane recovery project,
  98–99
  Lake Kivu, methane potential, 106–7

microhydropower, Nepal, 82

micro-restoration, 58–59

mini hydropower facility, 38–39

Ministry of New and Renewable Energy,
  India, 35

Mission Kakatiya, revival of traditional tanks,
  59–60
  M-Pesa and microfinance, 31–32

multipurpose dams
  bridge and dam, 21

dam for energy generation and irrigation,
  37–38
  flood control and irrigation, 38

"multi-utility tunnel" (MUT), 25–27
  Connaught Place (New Delhi), 26–27
  Putrajaya (Malaysia), 27
  Singapore MUT, 25
  SMART system (Kuala Lumpur), 27–29

Nairobi flooding jurisdictional questions, 184

Nashik Urban NEXUS program, 158–59

Nashik's Sustainable Urban Habitat Action
  Plan, 161

National Development and Reform
  Committee (NDRC), 43

National Hydro Power Corporation (NHPC), 37

National Merit Award for Sustainable Forestry
  Management, 170

National Water Supply and Drainage
  Board, 55

natural resource depletion, 8–9

Natural Step, sustainability platform, 14

New Delhi Energy Conservation Building
  Code (ECBC) cool roofing, 129–30

Next Generation Infrastructure (Brown), 5

Omnigrid Micropower Community Power
  program, 85–87, 253n30

On-site sanitary digester, 74, 198

Orr, David, 194, 211

Oujé-Bougoumou Cree First Nation village,
  93–94

"parafunctional" objects, 241n46

Parques Lineares de Fundo de Vale, 63

Paton, Charlie, 142, 144

Polavaram Dam, 59

politics of infrastructure projects, 182, 184
  Nairobi flooding jurisdictional questions, 183

polyculture (agriculture and aquaculture),
  41–43

population growth. *See* United Nations

Program 135 (Viet Nam), 151–52

Programme for Infrastructure Development in
  Africa (PIDA), 193–94
pumped storage hydro (PSH). *See* hydroelectric
  energy generation

rainwater harvesting, 52–53, 245n12
  Cistercian monks' monastic communities, 51
  India monsoon season, 55
  rain shadow, 245n4
  Sri Lanka, 48, 53–55, 75
Ramsar Convention, East Kolkata Wetlands
  (EKW), wetland of international
  importance, 72
Red River Delta mangrove restoration, 161
river crossings, multifunctions, 21. *See also*
  Jamuna Bridge; Khaju Bridge, Isfahan
road construction
  "multi-utility tunnel" (MUT) and, 27
  Rural Roads Improvement Program, tree
    planting, 131–32
  storm water management tunnel, 27–29
Robèrt, Karl-Henrik, 14
rope bridge rebuilding ritual, Peru, 147–49
Rural Development Ministry, India, tree
  planting, 265n77
rural electrification. *See* Omnigrid Micropower
  Community Power program
Rural Energy Development Program, 83–84
Rural Roads Improvement Program, tree
  planting, 131–32

Sacaton Pumped Storage Project, Tucson, 91
Safaricom (Kenya) and Commercial Bank of
  Africa accounts, 32
Sahara Forest Project, CSP and SWG synergy,
  143
Samra wastewater plant, 39
Santiago La Farfana Wastewater Treatment
  Plant, 74, 198
Sardar Sarovar Namada Nigam Limited
  (SSNNL) Canal, *36*, 37, 183
Saudi Arabia, Al Khafji utility-scale solar
  photovoltaic desalination plant, 141

Savory, Allan, 138
Savory Institute, 138
sea-level rise and coastal resilience, 116,
  161, 164
  Jakarta, seawall and land reclamation
    project, 119
Seawater Greenhouse (SWG), 142–43, 209,
  268n126
Seawater Greenhouse (SWG) pilot project, 143
silos, Infrastructural, 5, 159, 180, 183
  compartmentalizing infrastructure
    components, 9–11
social infrastructure, housing and
  transportation, *4*
soft-path solutions, 51, 58, 75, 196
  definition, 47–48
solar energy
  Belo Horizonte solar hot water, 99
  Betim solar thermal energy project, 92–93
  concentrated solar power (CSP) plant,
    143–44
  floating power stations, 37
  Jawaharlal Nehru National Solar Mission
    (India), 35–36
  Kagoshima Nanatsujima Mega Solar Power
    Plant (Japan), 37
  National Hydro Power Corporation (NHPC)
    (India), 37
  Omnigrid Micropower Community Power
    program, 85–87
  rooftop, 35, 53, 80–81
  rooftop water heating, 92, 201
  solar powered rickshaws, 110–11
Solecshaw, pedal and solar powered rickshaw,
  110–11
stepwells and baoli, 55
storage of energy. *See* hydroelectric energy
  generation
success, strategies for
  Alternative Heating Technologies, 201
  climate policies, 203–4
  collaboration for, 195–96
  decentralization, 206

success (cont.)
  green infrastructure with multiple benefits,
    197–98
  human role, 194–95
  Integrated Water Resources Management
    (IWRM) framework, 196–97
  low-carbon and renewable energy, 199–200
sustainability
  Agenda 21, 14
  Natural Step, sustainability platform, 14

Technologies for Economic Development
  (Lesotho), 74
telecom industry. *See also* Omnigrid
  Micropower Community Power program
  Digicel sale of excess capacity, 34–35
Tianjin Economic-Technological Development
  Area (TEDA), 43–46, 177, 244n82
tidal energy generation, tidal estuary power
  plant, parkland and transportation, 39
transparency and corruption, 191
Treatment Center for Solid Waste (CTRS), 100
Trung Son hydropower project, 38

UN Human Settlement Program Scroll of
  Honor, SMART system (Kuala Lumpur), 29
United Nations, 1
  Agenda 21, 14
  Clean Development Mechanism, 66–67, 201
  Millennium Development Goals, 78
  population growth, 6–7, 18
  Sustainable Development Goals, 78
  UN Habitat, 98
United Nations Department of Economic and
  Social Affairs, 6
United Nations Development Programme
  (UNDP), 32, 84
United Nations Environment Programme, 43
United Nations Industrial Development
  Organization, 39
United Nations Rural Energy Development
  Program, 83
Urban Land Institute (ULI), 187–88

urbanization, 7, 110. *See also* United Nations
  climate change and, 131, 137, 144
  problems of, 5–9, 98
  unrestricted, 61
utility companies
  American Society of Civil Engineers
    (ASCE), 189
  ANDE (Paraguay), 102
  CEMIG (Brazil), 100
  Electricité d'Haiti, 34
  Eletrobras (Brazil), 102
  GEMIG (Brazil), 92
  Lyonnaise des Eaux Casablanca
    (Morocco), 157
  regulators of, 189–90

Vatican statement on climate change, 115
Volta Redonda Eco-oil Program, 112

waste energy, Digicel sale of excess capacity,
  34–35
waste management, 201. *See also* wastewater
  treatment
  Belo Horizonte's waste pickers, 166–68
  East Kolkata Wetlands (EKW), 72
  Energy-from-waste (EFW), 260n136
  waste-for-transit privileges, 63
wastewater energy generation
  underutilized wastewater discharge, 39
  wastewater and solid waste treatment
    anaerobic digester, 96
wastewater treatment, 69–70, 73, 75, 250n91.
  See also biogas recovery; East Kolkata
  Wetlands (EKW)
  Diokoul and Surrounding Districts Sanitation
    Scheme, 164–66
  Jordan Valley Reclaimed Water Project
    (RWP), 137
  Santiago La Farfana Wastewater Treatment
    Plant, 74
water management, 47, 197–99, 204–5.
  See also hydroelectric energy generation;
  multipurpose dams; rainwater harvesting

Angkor Wat, Cambodia infrastructure model, 132
Cistercian monks' monastic communities, 51, *52*
hydrologic cycle management, 144–45
Lufumbu community water management project, 172–73
Nashik Urban NEXUS program, 158–59
raised field cultivation and lake-based development, 22,
reforestation impact in Ethiopia, 67–69
seesaw water pumps in Las Gaviotas, 32, *33*
water management, traditional
Bangladesh, rice straw into floating gardens (baira), 122
Chinampas in pre-Columbian Mesoamerica, 22–25
Khmer empire hydraulic network, 116–18
Lima terraces and stone canals, 133–34, *135*

Maner subbasin, 55–59
Sri Lanka ancient irrigation system, 48–51
water quality, 3, 66
water treatment plant damage, Liberia, 9
Waterbuurt, floating community on Lake IJ, 124
wind and renewable investments, Uruguay, 113
wind energy generation, 3, 80, 88–90
Jepírachi project, pilot wind farm, 154–56
wind-powered PSH system, 89–91
wind-powered PSH system funding, 91
World Bank, 9, 31, 108, 152–53
BioCarbon Fund, 66–67
Carbon Community Fund, 156
Carbon Finance Unit, 154
project funding, 19, 38, 54, 61, 84, 119, 150
Water and Sanitation Program, 66
World Economic and Social Survey 2013 (United Nations), 6